Quantum Physics for Beginners

Revealing the Secrets of Quantum Physics by Learning Simple and Useful Notions

Elvin L. Thomas

© Copyright 2021 by Elvin L. Thomas - All rights reserved.

This document is geared towards providing exact and reliable information in regards to the topic and issue covered. The publication is sold with the idea that the publisher is not required to render accounting, officially permitted, or otherwise, qualified services. If advice is necessary, legal or professional, a practiced individual in the profession should be ordered.

- From a Declaration of Principles which was accepted and approved equally by a Committee of the American Bar Association and a Committee of Publishers and Associations.

In no way is it legal to reproduce, duplicate, or transmit any part of this document in either electronic means or in printed format. Recording of this publication is strictly prohibited and any storage of this document is not allowed unless with written permission from the publisher. All rights reserved.

The information provided herein is stated to be truthful and consistent, in that any liability, in terms of inattention or otherwise, by any usage or abuse of any policies, processes, or directions contained within is the solitary and utter responsibility of the recipient reader. Under no circumstances will any legal responsibility or blame be held against the publisher for any reparation, damages, or monetary loss due to the information herein, either directly or indirectly.

Respective authors own all copyrights not held by the publisher.

The information herein is offered for informational purposes solely, and is universal as so. The presentation of the information is without contract or any type of guarantee assurance.

The trademarks that are used are without any consent, and the publication of the trademark is without permission or backing by the trademark owner. All trademarks and brands within this book are for clarifying purposes only and are the owned by the owners themselves, not affiliated with this document.

Table of Contents

Introduction 6

Chapter 1- What is Quantum Physics? 9
1.1 Getting Familiar 9
1.2 The Basics 13

Chapter 2- Max Planck - The Father Of Quantum Physics 22
2.1 Early Years 23
2.2 Contributions and Achievements in Physics 25
2.3 Character 29

Chapter 3- Einstein's Theory of Relativity 32
3.1 Unifying Time and Space 33
3.2 Unifying Energy and Mass 36
3.3 Proof from Experiments 38
3.4 Applications 42

Chapter 4- Schrodinger's Cat 45
4.1 What Counts as a Schrodinger's Cat? 47
4.2 What Schrodinger Really Wanted To Say? 51

Chapter 5- The Double slit Experiment 55
5.1 The Experiment 55
5.2 Do Atoms Passing Across A Double-Slit Know They're Being Watched? 63

Chapter 6- Development of the Quantum Concept 68
6.1 Matter Theories at the End of the 19th Century 71
6.2 The Issue of Black Body Radiation and Max Planck 74
6.3 Light Quantum and Albert Einstein 78
6.4 Duality in Waves and Particles 83
6.5 Atomic Spectra and Niels Bohr 85

6.6 Matter Waves of Schrodinger and de Broglie *86*

6.7 New Quantum Theory *91*

Chapter 7- Observing a Quantum System **99**

7.1 Observation Affects Reality *99*

7.2 What Does Quantum Theory Have to Say About Reality? *102*

Chapter 8- Quantum Leaps **108**

8.1 Reasons *108*

8.2 Controlling Quantum Leaps *113*

Chapter 9- Quantum Physics and Health **119**

Conclusion **128**

Introduction

This book you are about to read is my first job and to be able to achieve my goal, I need your help. Thanks to your positive review I can appear among the top positions, and therefore have more visibility and all this will be your merit.

Thanks in advance

Quantum physics is the field of physics used to deal with submicroscopic entities. Aspects in quantum mechanics are as mysterious and peculiar as aspects of relativity since certain entities are smaller than we can detect freely using our senses and must be viewed with the help of instruments.

Atoms and their substructure are well-known representations of objects that involve a thorough understanding of quantum mechanics. Classical physics theories exist for some of their properties, such as isolated electron shells. We envision isolated "electron clouds" around the nucleus in quantum mechanics.

We are acquainted with certain areas of quantum mechanics. We assume that matter is made up of atoms, the simplest form of an element, which comes together to make molecules, the simplest form of a compound. Although we cannot see single water molecules in a sea, we are mindful that this is due to the sea's large size and a large number of molecules. When we talk about atoms, we usually claim that electrons orbit them in isolated shells around a small nucleus, which is made up of many smaller particles such as protons and neutrons. We also know the electric charge is generated almost completely by protons and electrons in small units. We don't note individual charges in current flowing through a light-bulb as its charges are too tiny and abundant in the macroscopic conditions, just like we don't see individual water molecules in a sea.

Creating Links

Classical physics is a fair approximation of current physics. Quantum physics is appropriate in general and should be used instead of classical physics to define smaller particles, like atoms.

Molecules, atoms, and basic proton and electron charges are all quantized physical entities, meaning they only exist in distinct values and do not exist in all possible values. Continuous is the opposite of quantized. A quarter of an atom, a fraction of a proton's charge, or 14-1/3 cents, for instance, are not possible. Everything is instead made up of integral multiples of these structural elements. Quantum mechanics is a field of physics concerned with tiny particles and the quantification of different entities such as angular momentum and energy. Quantum physics, including classical physics, has many subfields, including mechanics and the analysis of electromagnetic powers. Quantum mechanics resemble classical physics in the classical limit (slow, larger particles), according to the correspondence principle.

Chapter 1- What is Quantum Physics?

What is quantum physics, and how does it work? Simply put, physics is the simplest explanation we have of the complexity of the objects that make matter and the forces by which they associate, and it describes how everything happens.

1.1 Getting Familiar

Quantum physics discusses how atoms work, and thus how biology and chemistry work. Quantum physics is needed to understand how electrons travel through a circuit board, how photons of light are converted to current in solar panels or enhance themselves in a laser, or also how the sun continues to burn.

Here is where the challenge – and, for enthusiastic physicists, the excitement – begins. To start with, no particular quantum theory exists. There are quantum mechanics, the fundamental mathematical mechanism that underpins everything, established by Niels Bohr, Erwin Schrödinger, Werner Heisenberg, and others in the 1920s. It describes basic things like the shift in direction or momentum of an individual particle over time.

Quantum mechanics should be coupled with other aspects of physics, most specifically Albert Einstein's theory of relativity, which describes what occurs as objects travel quickly, to construct quantum field theories, which describe how things function in the physical universe.

Three separate quantum field theories are used to describe three of the four basic forces that interfere with matter: electromagnetism, which describes how atoms keep together; the powerful nuclear force, explaining the structure of the nucleus at the center of the atom; and the weak nuclear force, explaining why certain atoms decay radioactively.

All three hypotheses have been thrown together in shambles known as the "standard model" of particle physics for the last five decades or so. Despite the fact that it seems to be bound together with adhesive tape, this model is the most fully checked portrayal of matter's fundamental workings ever invented. Its grandest success came in 2012 with the detection of the Higgs boson, the particle that provides all other fundamental particles their mass and whose presence had been predicted as far back as 1964 using quantum field theories.

The findings of experimentation at high-energy particle smashers, like CERN's (LHC) Large Hadron Collider, in which the Higgs boson was found, which investigate the matter at its tiniest scales, are well described by conventional quantum field concepts. However and if you'd like to know how things occur in much less abstruse situations – how electrons travel or don't travel in solids and so make an object an insulator a metal, or a semiconductor, for instance – things get much more complicated.

Image: The Large Hadron Collider (LHC) at CERN.

The billions of experiences in these crowded settings necessitate the creation of "efficient field theories" that brush over some of the grim information. Many crucial questions in solid-state physics remain unanswered due to the difficulties of building certain theories, such as why certain materials are superconductors at cold temperatures, enabling current to flow without resistance, and also why we cannot get them to conduct at normal temperature.

But there's a big quantum uncertainty lurking under all of these practical issues. Quantum mechanics predicts very peculiar ideas regarding how matter functions at a fundamental level that is entirely in conflict with how things appear to function in the physical universe. Quantum particles can act like particles when they are in a particular position, or they can behave like waves when they are scattered across space or in several positions at the same time. How they look seems to be determined by how we quantify them, yet until we measure them, they appear to have no specific properties at all, raising a profound conundrum regarding the existence and basic reality.

This ambiguity contributes to interesting paradoxes like Schrödinger's cat, in which a cat is both dead and alive due to an unknown quantum mechanism. Quantum particles often tend to be able to impact each other immediately, even though they are separated by a wide distance. Entanglement, or "spooky behavior at a distance," is the name assigned to this perplexing process by Einstein. We are totally unfamiliar with quantum forces, but they are at the core of new innovations like quantum computing and quantum cryptography.

Nobody understands what it all implies. Some argue that we must simply agree that quantum physics describes the material universe in ways that are incompatible with our understanding of the broader, "classical" world. Others believe there must be a stronger, more adaptive theory that we haven't discovered yet.

There are many big problems in all of this. For starters, quantum concepts have so far been unable to understand a fourth central force in existence. Einstein's theory of relativity, a distinctly non-quantum theory that does not even include particles, continues to govern gravity. Efforts to put gravity underneath the quantum framework and thereby clarify all of basic physics inside a single "theory of everything" have failed.

Additionally, cosmological calculations show that dark energy and dark matter make up about 95% of the cosmos, substances about which there is presently no understanding inside the conventional model, and dilemmas like the extent of quantum mechanics' position in the messy workings of existence remain unsolved. The universe is quantum in several ways, but whether quantum mechanics is the final word on the subject is an unanswered issue.

1.2 The Basics

Quantum physics is mainly just daunting at the start. Though physicists who work with it regularly find it odd and counter-intuitive. It is not, though, nonsensical. There are six main principles of quantum physics that you can consider while you're reading anything about it. If you do so, you'll find quantum mechanics far simpler to learn.

Waves make up everything, including particles.

There are several ways to begin a topic like this, and this is one of the best: anything in the world has both a wave and a particle structure at the very same time.

Obviously, representing actual things as both waves and particles is inherently a little messy. Quantum mechanics describes the third group of entities that share several characteristics of waves (a distinctive wavelength and frequency, some scattered through space) and some characteristics of particles (they're normally countable and may be localized to some extent). This has sparked a vigorous discussion in the physics community on whether this is acceptable to talk about light as a particle in introductory physics classes; not because there's some disagreement over whether the light has a particle structure, but because naming photons "particles" instead of "quantum field excitations" may contribute to some student misunderstandings.

The often perplexing terminology used by physicists to discuss quantum phenomena reflects this third nature of quantum properties. The Higgs boson was found as a particle at the LHC, but physicists often refer to the "Higgs field" as a delocalized object that occupies all of space. This occurs because it is more convenient to analyze Higgs field excitations in terms of particle-like properties in certain situations, such as collider simulations, whereas it is more feasible to discuss the mechanics in terms of encounters with a quantum field that fills the universe in other situations, such as general discussions of whether certain particles have some mass. It's just a different way of representing the same mathematical concept in a different language.

Quantum Mechanics Is Discrete

The term "quantum" derives from the Latin for "how many," and it refers to the idea that quantum models often contain anything that comes in distinct numbers. Quantum fields produce energy that is integer multiples of any fundamental energy. This is related to the wavelength and frequency of light, with short-wavelength, high-frequency light having broad characteristic energy and longer-wavelength, low-frequency light having a small limited characteristic energy.

However, in both instances, the cumulative energy found in a light field is an integer multiple of the energy— 1, 4, 18, 127 times— never an odd fraction like two-and-a-half or root of three. This property can also be found in atoms' distinct energy levels and solids' energy bands, in which some energy levels are allowed while others aren't. Because of the distinct nature of quantum physics, atomic clocks operate by utilizing the frequency of light correlated with a change between two permitted states of cesium to hold time at a pace that does not necessitate the much-discussed, newly discovered "leap second".

Ultra-precise spectroscopy may also be used to search for dark matter.

Also, fundamental quantum processes such as black-body radiation are inclined to involve continuous distributions, and this isn't often clear. But, digging through the equations reveals a granularity of the facts, which is a major part of what contributes to the theory's strangeness.

Quantum Physics involves probability

One of the most shocking and contentious facets of quantum physics is that the result of a controlled trial on a quantum system cannot be predicted with certainty. When physicists forecast the results of an experiment, they usually give a likelihood for discovering each of the specific potential outcomes, and similarities between hypothesis and experiment always require studying probability distributions from a large number of replicated experiments.

A "wave function" is a mathematical representation of a quantum field. There's a lot of controversy on what this wave function means, and it's split into two halves: those who believe the wave function is a true physical thing, and those who believe the wave function is simply a description of our understanding of the world ("epistemic" theories).

The likelihood of getting a result is not directly provided by the wave function in any category of the fundamental model but by the wave functions square (loosely talking; the wave function is a complex math object The likelihood of getting a result is not directly provided by the wave function in any category of the fundamental model but by the wave functions square (loosely talking; the wave function is a complex math object (meaning it includes imaginary numbers), and the function to get probability is marginally more involved. The "Born Rule," named after German physicist Max Born, who proposed it in 1926, is seen by others as an unattractive oddity. Some sections of the quantum foundations group are working hard to find a way to extract the Born rule from a more basic principle; so far, none of them have been completely accurate, but it's generating a lot of fascinating research.

This is also an element of the concept that allows particles to exist in several states at once. Only probability may be predicted because, before a prediction that decides a specific result, the mechanism being evaluated is in an indeterminate condition that arithmetically corresponds to a superposition of all possible outcomes of varying probabilities. If you think about this as the system being in all the states at once or only one uncertain state depends a lot on how you feel regarding ontic and epistemic structures, but both are restricted by the next point:

Quantum physics is a nonlocal science.

The last major contribution Einstein introduced to science was underappreciated, mostly as he was incorrect. Einstein made a clear mathematic statement about something that was troubling him for a certain time, a concept that we now term "entanglement,"

According to the EPR paper, quantum physics permitted the presence of systems in which measurements taken at widely spaced positions could be linked in such a way that the result of one was decided by the result of the other. They reasoned that this suggested that the calculation results had to be predetermined by a common factor since the alternatives would entail sending the outcome of one measurement to the position of the other at rates greater than the light speed. Therefore, quantum theory must be incomplete, a simple approximation to a broader principle.

For almost thirty years, this was thought to be an unusual footnote since there appeared to be no way to verify it, however in the 1960s, Irish scientist Named John Bell carried out the implications of the EPR paper in depth. Bell proved that quantum physics would predict greater associations between distant observations than any other conceivable theory of the kind favored by E, P, and R. This was studied experimentally by John Clauser in the 1970s, and Alain Aspect's early 1980s studies are generally regarded as conclusively demonstrating that these interconnected processes cannot be interpreted by any local secret variable theory.

The most popular explanation for this finding is that quantum physics is non-local, meaning that the effects of measurements taken at a specific position may be influenced by the properties of remote objects in ways that cannot be clarified using light-speed signals. This will not, however, allow for the transmission of knowledge at speeds faster than light, despite several efforts to do so using quantum non-locality. Refuting these has proved to be a remarkably fruitful endeavor. Quantum non-locality is also at the heart of the information challenge in evaporating black holes, as well as the "firewall" debate that has sparked a lot of recent debate. There are also some radical views involving a mathematical relation between wormholes and entangled particles, as defined in the EPR paper.

Quantum Physics Is Quite Small

Quantum mechanics has a reputation for being strange because its results are too dissimilar to our daily reality. This is because the results involved become smaller as things become bigger—if you want to view unmistakably quantum activity, you generally want to see particles behave like waves, and the wavelength reduces as momentum rises. The wavelength of a macroscopic entity such as a cat going through the room is so small that if you stretched it to the scale of the whole Solar System, the cat's wavelength will be around the size of a single atom inside the solar system.

This implies that quantum effects are often limited to the size of fundamental particles and atoms, where velocities and masses are tiny enough for wavelengths to become broad enough to detect directly. However, there is a concerted attempt in several ways to increase the scale of systems that exhibit quantum effects. There have even been reports that this could be achieved with suspended mirrors weighing several grams, which would be incredibly awesome.

Quantum mechanics isn't wizardry

The previous argument inevitably leads to this one: quantum mechanics is not magic, no matter how odd it may appear. By ordinary physics norms, the stuff it predicts is odd, but they're intensively bound by understood scientific laws and concepts.

So, if anyone confronts you with a "quantum" concept that seems too amazing to be true—free electricity, magical curing abilities, unimaginable space drives—it almost definitely is. That's not to say we can't use quantum mechanics to do extraordinary stuff—some there's pretty interesting physics in everyday technology—but those things remain well inside the confines of thermodynamics and common sense.

Chapter 2- Max Planck - The Father Of Quantum Physics

Max PLANCK (1858-1947)

Planck made numerous contributions to physics, and he is best known for being the founder of quantum theory. Much like Albert Einstein's theory of relativity reshaped our view of time and space, his theory expanded our knowledge of subatomic and atomic processes. Next to each other, they constitute the basic theories of 20th-century physics. Both have persuaded society to reconsider some of its most profoundly rooted scientific convictions, and both have culminated in military and industrial applications that impact every dimension of daily life.

2.1 Early Years

Planck's own life and career are greatly influenced by his family's long history of loyalty to state and church, academic achievement, corruptibility, idealism, conservatism, honesty, and kindness. Planck's father was assigned to the University of Munich when he was nine years old, and Planck enrolled in the city's famed Maximilian Gymnasium, where a coach, Hermann Müller, ignited his passion in math and physics. Planck, on the other hand, excelled in all topics, and when he graduated at the age of 17, he had a tough career choice. He eventually preferred physics over music or classical philology because he had come to the dispassionate opinion that physics was where he had the most originality. Despite this, music remained an important aspect of his life. He had the gift of perfect pitch and was an outstanding pianist who found peace and joy at the piano every day. Particularly in his advanced years, he embraced the outdoors, enjoying long walks every day and vacationing in the mountains to trek.

In the autumn of 1874, Planck enrolled at the University of Munich, but the physics professor gave him no support. Despite their greatness as scientists and researchers, he was disappointed by Gustav Kirchhoff's and Hermann Helmholtz's lectures during the period at the Berlin University. His analytical abilities, on the other hand, were brought into sharper focus as a product of his independent research. Having returned to Munich, he earned his doctorate at the exceptionally young age of 21 in 1879 (birth year of Einstein). He became a lecturer the next year after completing his qualifying dissertation in Munich. He was named associate professor at the University of Kiel in mid-1885, thanks to his dad's professional contacts. Following Kirchhoff's passing in 1889, Planck was assigned to the Berlin University, where he began to consider Helmholtz as a teacher and colleague. He was elevated to full Professor in 1892. He only had nine doctoral students in all, yet his lectures on all categories of theoretical physics have been reprinted several times and have had a major effect. He spent the remainder of his life in Berlin.

Planck said his decision to pursue science was based on his discovery...that the laws regulating the sequences of the perceptions we obtain from the environment around us agree with the laws of human reasoning; that, therefore, pure logic will allow us to obtain an understanding of the mechanisms of the universe. In other terms, he consciously wanted to become a physicist during a period when theoretical physics had not yet been known as a distinct discipline. However, he went on to claim that the presence of physical laws means that the "outside universe is something separate from man, something pure, and the search for the laws that refer to this fundamental appeared...as the most sublime empirical endeavor in life."

The law of conservation of energy, also known as the 1st law of thermodynamics, was the first example of an absolute of nature that greatly impressed Planck, even as a Gymnasium student. Later, during his undergraduate years, he became persuaded that the second concept of thermodynamics, the law of entropy, also was absolute universal law. The second rule was the focus of his doctoral dissertation in Munich, and it was at the heart of the experiments that contributed to the discovery of Planck's constant h.

2.2 Contributions and Achievements in Physics

A blackbody, as described by Kirchhoff, is an entity that reemits all radiant energy upon it, — in other words, it is a complete absorber and emitter of radiation. By the 1890s, numerous theoretical and experimental efforts had been made to calculate its spectral energy distribution—the spectrum showing how much radiant energy is released at different wavelengths for a defined blackbody temperature. Planck was especially drawn to a formula discovered by his partner Wilhelm Wien, and he tried to extract "Wien's rule" from the 2nd thermodynamics law. Other PTR colleagues had discovered definite signs that Wien's rule, though true at high frequencies, failed entirely at low frequencies.

Just before a conference of the German Physical Society, Planck heard of these findings. If Wien's rule persisted in the high-frequency area, he understood how the radiation entropy had to be mathematically based on its energy. He also saw how this dependency had to be in the low-frequency area if the experimental effects were to be replicated. Planck reasoned that he should attempt to integrate these two expressions as simple as possible, and then convert the effect into a calculation that relates its frequency to the energy of radiation.

The outcome, regarded as Planck's radiation rule, was universally praised as unmistakably right. Planck, on the other side, treated it as just a hypothesis, a "fortunate intuition." It had to be built from the first concepts if it was to be treated seriously. It was Planck's first priority, and by 14, 1900, he had accomplished it—albeit at a significant expense. Planck discovered that in order to accomplish his aim, he had to abandon one of his most valued beliefs: that the 2nd thermodynamic rule was an absolute nature law. Instead, he had no alternative but to follow Ludwig Boltzmann's understanding of the second law, which was that it was a mathematical law. Furthermore, Planck had to conclude that the oscillators that make up the blackbody and re-emit the energy occurring upon them cannot consume this energy constantly, but only in distinct quantities, in energy quantas; and that the only way to disperse these quanta, each carrying a level of energy h proportional to their frequency, across all of the oscillators existing in the blackbody. He used his theorem to measure the constant h, and also the so-called Boltzmann constant (the basic constant in statistical mechanics and kinetic theory), the electron charge, and Avogadro's number. Over time, physicists discovered that the microphysical universe, the realm of atomic dimensions, could not be defined in theory by ordinary classical mechanics since Planck's constant was not null but had a tiny but finite meaning. A major shift in physical science was on the horizon.

In other terms, Planck's idea of energy quanta ran counter to any previous physical theory. He was forced to implement it purely by the influence of his logic; he was a hesitant activist, as one scholar put it. Indeed, it took years for the far-reaching ramifications of Planck's accomplishment to be universally understood, and Einstein played a crucial role in this. In 1905, independently of Planck's work, Einstein proposed that radiant energy seemed to be made up of quanta (called photons later on), and in 1907, he demonstrated the generalization of the quantum principle by applying it to the temperature dependency of solid particular heats. The wave-particle duality was brought into physics by Einstein in 1909. Einstein and Planck were among the influential physicists who participated in the very first Solvay conference in Brussels in 1911. Henri

Poincaré was motivated by the discussions to provide statistical proof that Planck's radiation law involved the inclusion of quanta—a proof that won over James Jeans and some others to the quantum theory. Niels Bohr's hydrogen atom quantum theory led significantly to its establishment in 1913. Oddly, Planck became the last to argue for a transition back to classical theory, a position he eventually grew to reconsider rather than use as a way of fully reminding himself of the quantum theory's necessity. Despite the observation of the Compton Effect in 1922, resistance to Einstein's extreme light quantum theory of 1905 continued.

In 1900, Planck was 42 years when he produced the renowned breakthrough that gained him the Nobel Prize for Physics in 1918 and several other honors. It's not shocking that he didn't make any more important findings after that. Despite this, he managed to make significant contributions to diverse fields such as optics, statistical mechanics, thermodynamics, physical chemistry, and others. He was also the first well-known scientist to advocate for Einstein's theory of relativity. Planck and chemist Walther Nernst were involved in taking Einstein to Berlin in 1914, and after the war of 1919, steps were taken for Planck's favored pupil, Max von Laue, to visit Berlin as well. Once Planck retired in 1928, Erwin Schrödinger, the pioneer of wave mechanics, was selected as his heir. For a while, Berlin shone brightly as a hub of physics—until the ascension of Hitler in January 1933 put a shadow over the region.

2.3 Character

Planck's works became more metaphysical, aesthetic, and theological in nature as he grew older. He, like Schrödinger and Einstein, was absolutely opposed to the statistical, indeterministic worldview brought into physics following the introduction of quantum mechanics in 1926 by Bohr, Werner Heisenberg, Max Born, and others. Planck's strongest intuitions and convictions were incompatible with such a viewpoint. Planck claimed that the real world is an external force that occurs independently of man; unlike Bohr and his school, the observer, as well as the observed, are not closely coupled.

In 1912, Planck was appointed permanent secretary of the Prussian Academy of Sciences' physics and mathematics divisions, a position he retained until 1938; he also was head of the Kaiser Wilhelm Society (now known as Max Planck Society). These and other positions put Planck in a role of great power, — particularly among German physicists; his opinions and advice were rarely challenged. His power, on the other hand, came mostly from his own spiritual influence, not from the official positions he occupied. His honesty, sincerity, and intelligence were undeniable. Planck's overt approach to Hitler in an effort to undo Hitler's disastrous racist policies was entirely in character, as was his decision to stay in Germany throughout the Nazi era to retain everything he could of physics in Germany.

Planck was a guy of a strong will. He may not have survived the tragedies that occurred in his life after the age of 50 if he had become less stoic and had less moral and philosophical belief. After 20 years of being married, his 1st wife, Marie Merck passed in 1909, having left Planck with 2 sons and twin girls. Karl, the eldest sibling, was killed in battle in 1916. Margarete, one of his children, died during childbirth the next year, and Emma, his remaining daughter, died in the same manner in 1919. More disaster arose as a result of World War II. In 1944, bombings totally demolished Planck's home in Berlin. Worse, the younger sibling, Erwin, was involved in the assassination attempt on Hitler in 1944, and he suffered a horrifying death by the Gestapo in 1945. Planck's desire to live was broken by that callous gesture. Planck and his 2nd wife, Marga Hoesslin, who he had wedded in 1910 and with whom he had had 1 son, were taken to Göttingen by United States officers at the end of the war. He passed there in 1947, at the age of 89.

Planck showed that energy would take on the properties of physical matter in some circumstances through experimental observations. Energy, according to classical physics theory, is solely a persistent wave-like force that is unaffected by the properties of physical matter. According to Planck's principle, radiant energy consists of particle-like elements called "quanta." The hypothesis assisted in the interpretation of previously unknown natural phenomenon such as the action of heat within solids and the existence of atomic light absorption. Planck received the Nobel Prize for his thesis on blackbody radiation.

Others, like Albert Einstein, Louis Broglie, Niels Bohr, Erwin Schrodinger, and Paul Dirac, progressed Planck's principle and paved the way for the advancement of quantum mechanics—a scientific interpretation of quantum theory that asserts that energy may be both wave and matter based on such factors. Quantum mechanics takes a probability-based view of existence, in comparison to classical mechanics, which claims that the precise properties of items are quantifiable in theory. Modern science is based on the mixture of Einstein's theory of relativity and quantum mechanics.

Chapter 3- Einstein's Theory of Relativity

Albert Einstein presented his special relativity theory in 1905, which describes how to perceive motion between various inertial framings of reference — positions that travel at constant speed relative to one another.

Instead of referring to the ether as an arbitrary point of reference that determined what was going on, Einstein clarified that when two body moves at a steady pace, the relative motion between both the different bodies defines what is going on. Whether you and another astronaut, John, are traveling in separate spacecrafts and want to contrast your findings, the only thing that counts is how much you and John are flying relative to each other.

Just the special case of uniform motion is included in special relativity. Only if you're moving along a straight line at a steady pace would it justify the motion. Special relativity is no longer true as soon as you curve or accelerate — or do something else that alters the structure of the motion in some way. This is where Einstein's general theory of relativity falls in since it can describe any form of motion in general.

Einstein's theory was founded on two fundamental ideas:

- **The relativity principle:** Even with events traveling in inertial (constant speed reference frames, the rules of physics should not alter.

- **The light speed principle:** The speed of light is constant for all witnesses, regardless of their position in relation to the light source. (Physicists use the symbol c to represent this speed.)

Einstein's findings were brilliant because he looked at the observations and concluded the results were right. This seemed to be the polar opposite of what most scientists were doing. He believed the results were right and the hypothesis had failed, rather than the theory being correct and the experiments failing. Physicists began looking for the elusive material known as ether in the late 1800s, claiming it to be the means by which light waves might travel. According to Einstein, the assumption in ether has messed it up by adding a medium that induced certain rules of physics to function differently based on how the observer traveled compared to the ether. Einstein simply dismissed the ether and believed that the rules of science, like light speed, existed similarly regardless of how you move — precisely as studies and mathematics proved!

3.1 Unifying Time and Space

The theory of special relativity, developed by Albert Einstein, established a fundamental relation between space and time. There are three space axes/dimensions — left/right, up/down, backward/forward — and one-time dimension in the universe. The space-time continuum refers to this 4-dimensional space.

If you travel quickly enough across space, your observations regarding space and time can vary somewhat from those of other people traveling at various speeds.

Understanding the thinking experiment illustrated in this figure will help you visualize this. Assume you're on a spacecraft, and you're carrying a laser that fires a ray of light straight up, hitting a mirror you've hung from the ceiling. The beam of light then returns to the ground and collides with a detector.

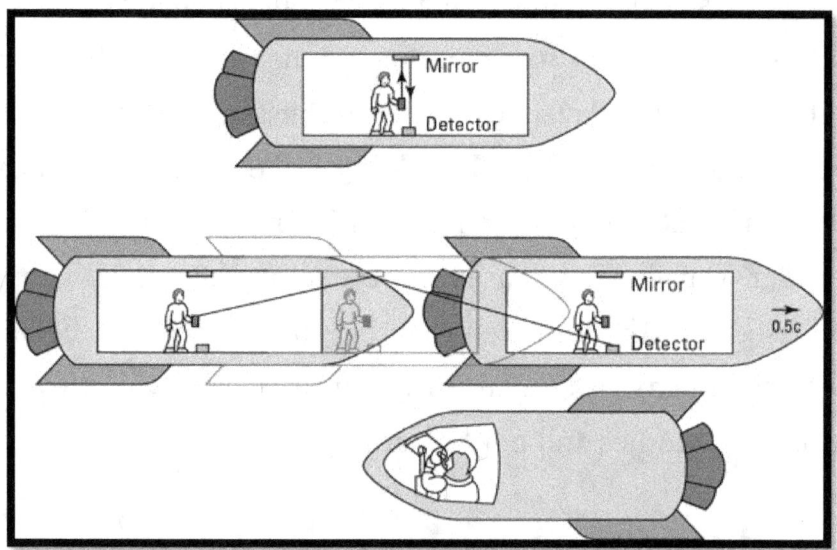

(Top image) A ray of light ascends, bounces off the mirror, then descends straight. (Bottom image) John notices the beam moving in a diagonal direction.

The spacecraft, on the other hand, maintains a steady speed of half of light speed (0.5c, as scientists would say). This, as per

Einstein, makes no sense to you when you can't feel you're traveling. It will be a different matter if astronaut John was spying on you, as seen at the bottom of the picture.

John will see the beam of light go up a diagonal path, touch the mirror, and then down a diagonal trajectory before hitting the detector. To put it another way, you and John will see separate light paths, and those pathways aren't even the same in length. This implies that the period it takes for the ray to travel from the source to the mirror and then to the detector will vary for you and John in order for you to settle on the speed of light.

Time dilation is a phenomenon in which time tends to move slowly on a fast-moving ship than it does on Earth.

As odd as it can be, this illustration (among many others) shows how time and space are inextricably related in Einstein's theory of relativity. As Lorentz transition equations are used, the light speed for all observers becomes completely the same.

Since this peculiar behavior in time and space is only apparent while flying at light speed, no one has ever encountered it before. Experiments performed after Einstein's invention have shown that it is real — space and time are interpreted differently for particles traveling at the speed of light, just as Einstein expected.

3.2 Unifying Energy and Mass

The most popular work of Einstein's existence also dates to 1905, when he used the theories from his relativity study to develop the formula E=mc2, which expresses the connection between energy (E) and mass (m).

In summary, Einstein discovered that the mass of the object grew as it reached the light speed, c. The target becomes heavier when it moves faster. The object's energy and mass would all be infinite if it could simply travel at c. Since a heavier mass is more difficult to accelerate, getting the particle to c is impractical.

After a ten-year effort to incorporate acceleration into the theory, Einstein revealed his general theory of relativity in 1915. He discovered that large forces create a distortion of time and space, which is experienced as gravity.

Gravitational pull

The power of attraction between two points is defined as "gravity." When Sir Isaac Newton proposed his 3 laws of motion, he quantified the gravitational attraction between 2 objects. The pulling force between 2 bodies is proportional to their mass and the distance between them. Your center of mass is pulling at the Earth, even though the Earth's center is pulling you against it (helping to keep you firmly planted on the ground). However, the huger object scarcely feels your pull, while the much smaller mass is deeply embedded thanks to the same power. However, Newton's laws presume that gravity is a natural phenomenon that will operate over a long distance.

Albert Einstein established that the rules of physics are the same across all non-accelerating observers in his principle of special relativity, and he demonstrated that the speed of light inside a vacuum is the same regardless of the velocity at which an observer moves. And so, he discovered that time and space were intertwined to form the space-time continuum. Things that occurred at the very same time for one observer can take place at different intervals for another observer.

Einstein discovered that large forces created a distortion of space and time when he carried out the calculations for his general theory of relativity. Try placing a wide body in the middle of a trampoline. The cloth would dimple as the body pressed back onto it. A marble rolling along the edge will spiral downward into the body, just like the gravitational force of a planet pulls on objects in space.

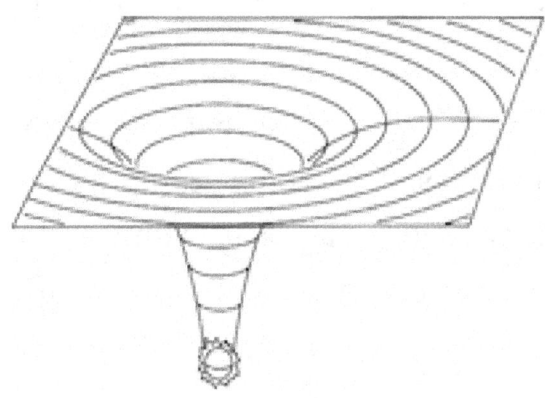

3.3 Proof from Experiments

Several of the phenomenon expected by time-space warping have been observed, even though instruments cannot see or quantify it.

Gravitational lensing can be seen in Einstein's Cross. (NASA and the European Space Agency (ESA) contributed to this image.)

Gravitational lensing occurs as light is bent through a large point, like a black hole, allowing it to serve as a lens to the objects behind it. This technique is commonly used by astronomers to research celestial bodies that are hidden behind large objects.

Einstein Cross, which is a quasar inside the constellation of pegasus, is a good demonstration of lensing gravity in action. It is approximately 7 billion huge light-years distant from Earth and is situated behind a galaxy 400 million light-years out. Since the galaxy's strong force twists the light emanating from the quasar, four photographs of the quasar emerge around the galaxy.

Scientists have been able to see some pretty interesting stuff thanks to gravitational lensing, but what they've seen through the lens has stayed somewhat stagnant until recently. Scientists were able to detect a supernova four times when it was magnified by a large galaxy and light moving through the lens follows a separate direction, each traveling for a certain period of time.

Another fascinating discovery was made by NASA's Kepler telescope, which discovered a white dwarf (dead star) circling a red dwarf. Despite its greater size, the white dwarf has a much smaller radius than a red dwarf.

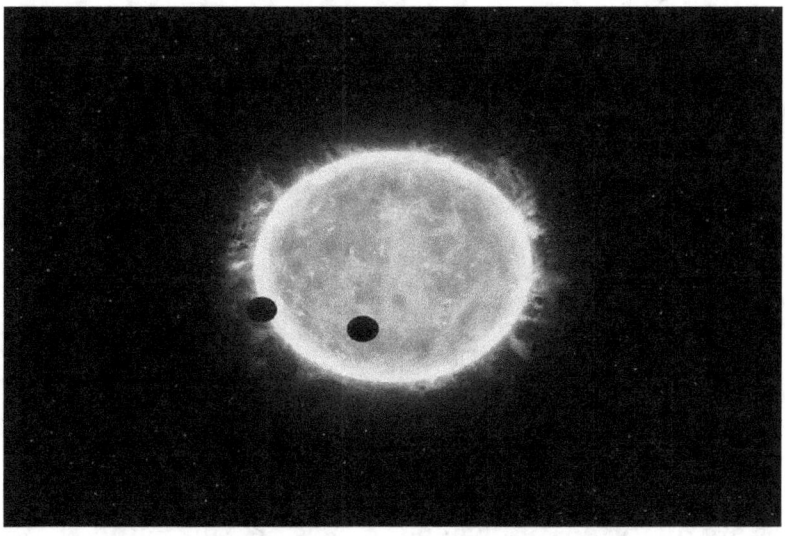

Image: Planets moving in front of a red dwarf.

In a tweet, a scientist said, the method is equal to detecting an insect on a tree 5,000 kilometers away, approximately the gap between Los Angeles and New York City."

Changes in Mercury's Orbit: Owing to the curving of time-space around the sun, Mercury's orbit is steadily changing over time. It could also crash with Earth in a couple of billion years.

Space-time frame dragging through rotating bodies: The rotation of a large mass, like the Earth, can twist and bend the space and time around itself. Gravity Probe B (GP-B) was introduced by NASA in 2004. The satellite's accurate calibration allowed the axes of the gyroscopes within to drift quite marginally over time, confirming Einstein's hypothesis.

In a tweet, GP-B chief scientist Francis Everitt of Stanford said to view the Earth as though it were submerged in honey

The honey swirls as the earth rotates, and the same is true in time and space. GP-B has verified two of Einstein's most profound theories, with far-reaching significance for astrophysics science.

Within a gravitational field, an object's electromagnetic radiation is spread out somewhat, resulting in gravitational redshift. Consider the sound waves produced by an emergency car's siren: when the vehicle reaches an observer, the sound waves are distorted, but when it drives on, the sound waves are spread out, or redshifted. The Doppler Effect is a phenomenon that happens with visible light at all wavelengths. 2 scientists at Harvard, fired gamma-beams of radioactive iron above the side of a tower in 1959 and discovered that they were minutely less than their normal frequency owing to gravity disturbances.

Gravitational waves: Violent phenomena, like the crashing of 2 black holes, are believed to be capable of causing gravitational waves, which are ripples in space and time.

By using Background Imaging of Cosmic Extragalactic Polarization (BICEP2) telescope in deep Antarctica, scientists declared in 2014 that they had observed gravitational waves remaining from the Big Bang. These waves are believed to be present in cosmic microwave history. Further investigation found, though, that their data had been tainted by debris. The Laser Interferometer Gravitational-Wave Observatory (LIGO) reported that it had spotted proof of alert signals left by gravitational waves.

In 2015, LIGO detected the first reported gravitational wave. The instruments had just been updated and were in the course of being configured before going live. As per LIGO spokeswoman Gabriela Gonzalez, the first discovery was so big that it required the department several months of analysis to persuade everyone that it was a true indication and not a bug.

A 2nd signal was discovered in December of the same year, it was accompanied by the declaration of a third contender. Though the first 2 signals are most certainly astrophysical—Gonzalez claims there is a rare chance they're not—the third only has an 85% chance of becoming a gravitational wave.

The proof for black hole pairs spiraling downward and colliding is provided by the two detections taken together. Gonzalez believes that as time goes by, LIGO and other future devices can identify further gravitational waves.

Gonzalez said, "We have assessed general relativity, and it has completed the test."

3.4 Applications

Many modern developments have been made possible thanks to Einstein's principle of relativity. Without relativity, there will be no global positioning system, radar cannons, or cathode ray television, among other things. T televisions with cathode ray tubes (CRTs) produce images by firing electron particles at a film of phosphorous. These electron particles are accelerated to near-light speeds of 20-30% of the light speed. Recall from special relativity that when a particle reaches light speed, the amount of energy available to propel it increases. The electrons are positioned in the right configuration on the frame by magnets in the television. They must compensate for the relativistic influence on these electrons, otherwise, the image would be distorted. CRT televisions, the precursors of plasma, were made possible thanks to Einstein.

The radar gun is another modern technology that has a connection to Einstein's theory of relativity. Radar weapons are being used in the military, competitive athletics, and, even, speed sensors. A radio frequency emitter and a detector make up a radar gun. A cop aims this gun at a moving vehicle. Radio waves are emitted by the radar gun. These waves re-enter the detector after bouncing off the moving driver. The radar gun will measure the speed of a moving car based on the sum of wavelength change caused by the Doppler Effect, which is mentioned in special relativity. So, thank Einstein for getting your speeding tickets.

The GPS precision is dependent on the principle of relativity. The GPS constellation comprises 24 satellites orbiting the Earth. These satellites are approximately 20,000 kilometers above the surface and traveling at a pace of 4 kilometers per second. Four signals generated by four satellites are processed by GPS antennas, such as the one in your mobile's navigation device. They use the signals to triangulate your position by solving four separate linear equations. The distance traveled by the wave (the length from you to the satellite) is equal to the velocity of the wave (light speed) times the time it required the wave to hit you. GPS satellites should maintain extremely precise time for these calculations to function correctly. They make use of atomic clocks which are calibrated for both general and special relativity effects regularly. .The time dilation caused by the relative speed of the earth and the satellites can delay the GPS clock. Gravitational time dilation, on the other hand, is exacerbated by satellite clocks becoming farther away from the Earth and hence in less curved space and time than those on Earth's land. The mistakes would be large if these consequences were not taken into account; as high as 12 km in just one day. Einstein's discovery influenced electron beam machines as well (EBM). EBM is a non - traditional machining technique that uses a beam of high-speed electrons to conduct cutting operations. It was patented in 1952. The electron particles are generated in an electron beam gun and then accelerated to 60-80% of light speed via an anode. For speeds approaching the light speed, the electron kinetic energy must be measured using special relativity calculations. The kinetic energies of these high-energy electrons are converted to thermal energy as they collide with the work piece. The material is melted or vaporized by this force, resulting in the required cut. Quick, high-precision machining of a range of parts for the automobile, aerospace, medical, and other sectors has been made possible by EBM. The fundamentals of Einstein's general and special theories of relativity have given a clear explanation of one of the most influential theories of the twentieth century. The theory of relativity clarified previously unknown experimental findings, paved the way for modern scientific breakthroughs, and allowed the creation of several modern technologies.

Chapter 4- Schrodinger's Cat

One of the founding fathers of classical physics, Erwin Schrödinger is known for a variety of significant contributions to science, most notably the Schrödinger equation, by which he was awarded the Nobel Prize in 1933.

Erwin Schrödinger's research in quantum mechanics solidified his reputation inside the field of physics. His furry paradox hypothetical scenario has been a popular culture classic.

A Geiger detector, a hammer, a vial of venom, and a radioactive agent are all put in a metal container with a kitten. As the radioactive material decays, the Geiger detector senses it and activates the hammer, which releases the toxin, killing the animal. Radioactive decay is a spontaneous phenomenon with no means of predicting when it will occur. The atom, according to scientists, is in a configuration called superposition, in which it is simultaneously not decayed and decayed at the same moment.

An outsider has no way of knowing if the kitten is dead or alive before the container is opened, since the cat's existence is inextricably linked to whether or not the atom has decayed, and the cat will be "dead and alive... in equal sections" before it is observed, as Schrödinger described it.

In other terms, until the container becomes opened, the cat's condition is unclear, and as a result, the cat is thought to be both dead and alive till it is observed.

When you bring the kitten in the container and there is no way of knowing what the kitten is doing, you're [likely] liable to be incorrect if you want to create assumptions based on your knowledge of the cat's state. On the other side, you'll be right if you think it's in a blend with all the possible situations."

An observer might decide if the cat was dead or alive only from looking at it because the cat's "superposition"—the notion that it was in all states—would crumble into either the awareness that "the cat is dead" or "the cat is alive," yet not both.

Schrödinger created the paradox to explain in classical physics regarding the existence of wave particles.

What we found around the 1900s was that very small particles did not follow Newton's Laws. As a result, the laws that regulate the movement of a ball, a human, or a vehicle can't be used to describe how a photon or an electron functions."

The concept of a wave function is at the core of quantum mechanics and is used to understand how subatomic particles including protons and electrons behave. A wave function defines all of the potential states of such objects, including momentum, energy, and location.

The wave function is a composite of all the potential wave functions. A particle's wave function states that it has a chance of being in all of the possible positions. However, without seeing it, you can't assume you believe it's in a certain location. Until we observe it and determine where it is, an electron placed around the nucleus may have all of the permitted states or locations. Schrödinger was explaining this with his paradox.

4.1 What Counts as a Schrodinger's Cat?

This is a powerfully strange concept, and the sight of something large being scattered through two states at the same time has been enthusiastically embraced by popular media. To name only three latest Google News search results, references to Schrödinger's cat can be found in political satire, art critique, and commentary of sports.

The concept of "Schrödinger cat" has sparked a lot of interest in physics, with researchers delving further into the phenomena.

It's worth contemplating what makes a "Schrödinger cat" in an actual experiment.

A single quantum device in a superposition of different states is the easiest scenario. P scientists have been at it for more than 25 years. They got a single trapped ion and put it in a superposition condition that enabled it to be in two places at the same time, on separate ends of the ion trap.

This experiment, as well as several others since explicitly demonstrates the kind of two-states-at-once action that is fundamental to most common Schrödinger pet glosses. To certain people, it should not even matter because it's only one particle. The "ridiculous scenario" that Schrödinger imagined included a massive structure that was in several states.

Putting a large number of atoms into a similar superposition condition is the next step of the experimental "cat state" statements. The superposition of counter-clockwise and clockwise currents along a small superconducting circle was first shown in a paper published in 2000. Millions of electrons were involved, all performing the same uncertain thing.

However, several physicists believe that this isn't a "real" Schrödinger cat since each one of those specific particles was doing its own act, essentially independent of others. In certain ways, it's really a bunch of single-atom versions piled above one another.

A big composite object put in a superposition condition is what those scientists require before they admit the existence of a true Schrödinger cat. Such that, they would like a series of thousands of particles joined together in a single entity being in a superposition condition collectively, with all of their bonds intact.

It is a daunting task, but incremental progress has been made utilizing structures that are small on a human level but enormous in terms of single atoms. A whirlwind of recent reports regarding the newest of these developments, concerning small "drumheads" put in superpositions of various vibrational conditions, is the immediate trigger of the message. The membranes in issue are thinner than a mere cat fur, but the phases are aggregate phases with all the thousands of atoms that make up the membrane flowing together, giving it the consistency of "largeness" that Schrödinger's "very absurd" scenario needs.

The most difficult aspect of these studies is demonstrating that the condition of the membrane is truly indeterminate—distinguishing between "definitely alive or definitely dead" also "both dead and alive at the very same moment," like in many other quantum phenomena. This necessitates an experiment through which the two pieces are recombined in such a manner that they interfere, resulting in an oscillation in the probability of different outcomes. The larger the system, the more difficult it is to manage enough to test in a repeatable manner, and this is the main obstacle that physicists face.

However, there is another way to look at Schrödinger's imaginary cat's meaning: it's not just about the cat. If you check at the initial paper that began this entire thing, you might make a good case that his true interest is the entanglement between both the atom and the cat. The paper is a study of quantum mechanics as it was understood in 1935, as well as a metaphysical reflection of Schrödinger's discontent with it.

The argument of the paper's cat illustration isn't so much that the cat is in an insanely indeterminate situation as it is that its condition is inextricably linked to the atom. It's a thinking experiment aiming at the idea that like certain iterations of the Copenhagen Interpretation tried, you might make a total separation between large objects that act classically and microscopic things that behave quantum-mechanically. The argument made by Schrödinger was that this distinction is untenable: you can easily set up tests in which the condition of big things is completely dependent on the state of small things, forcing science to deal with what's "actually" going on rather than dismissing it as unobservable and therefore meaningless. It's all connected to the problem of observers and interpretations.

In that context, all of the previously described "cat state" tests could "count," since they all require any measurement apparatus being entangled with a micro quantum entity in an undefined state. Of necessity, this brings with it a slew of metaphysical and interpretive luggage.

Whichever way you look at it, Schrödinger's cat has undeniably acted as a source of motivation for physicists and philosophers alike. The disputes about what it means to be in a "Schrödinger cat condition" aren't going anywhere anytime soon, and experimental advances toward "cat states" in larger and larger objects will continue to propel this further into new fields.

4.2 What Schrodinger Really Wanted To Say?

Since the quantum theory was first formulated in the early twentieth century, physicists have struggled over how the laws of quantum mechanics translate into the seemingly very different principles of classical mechanics — where particles have well-defined properties, locations, and directions. Is there a basic distinction between massive classical and tiny quantum objects? Schrödinger's thought experiment famously highlighted the dilemma of the quantum-classical transition.

The unfortunate kitten is a misunderstood creature. The point of Schrödinger's argument was not, as is often assumed, the inherent inconsistency of quantum mechanics when extrapolated to a larger scale. The pet was the result of communication between Erwin Schrödinger and Einstein, following Einstein's criticism of the Danish scientist Niels Bohr's understanding of quantum physics.

As explained before; Quantum mechanics, according to Bohr, seems to compel one to assume that the properties of quantum particles such as electrons don't have definite values until they are measured. It seems absurd to Albert Einstein that any aspect of life needs our deliberate attention to come into existence. In 1935, he proposed a thought experiment with 2 fellow classmates, Nathan Rosen and Boris Podolsky, that appeared to rule out that explanation. The 3of them (whose study is now known as EPR) discovered that particles can be formed in states that must be associated with one another, in the respect that any of them has a specific value for a property, other one should have a specific value as well. When two electrons have the properties of spin, one spin may point "up," whereas the other electron's spin looks "down."

If Bohr was correct, and the exact positions of the spins remain unknown before they are measured, so the similarity of the 2 spins implies that calculating one of them automatically fixes the direction of the other, no matter how distant the particle is. This ostensible relation was dubbed "spooky behavior at a distance" by Einstein. However, such a phenomenon may be impractical due to Einstein's special relativity principle, which states that no force will travel quicker than light speed.

Experimentally and technically, whether there is a limit to the scale of structures that can be intertwined with one another is an unanswered issue.

The correlation here between particles was dubbed "entanglement" by Schrödinger. Since the 1970s, experiments have demonstrated that it is a true quantum concept. However, this doesn't imply that quantum particles will immediately manipulate each other through space as a result of Einstein's strange behavior. It's more accurate to conclude that an individual particle's quantum characteristics aren't always determined at a single fixed location in space, they may also be "nonlocal": fully stated only in comparison to another particle somewhere, in a way that seems to contradict our innate understanding of time and space.

Schrödinger's cat emerged from his musings on EPR entanglement's intricacies. If we visualize blowing entanglement up to a daily scale, Schrödinger tried to demonstrate how Bohr's idea that nothing is set before it is measured would contribute to logical insanity. or In his thinking experiment, the unfortunate cat is trapped in a locked container with a bottle of poisonous substance that can be torn open by a device that connects it to — or entangles it with — a quantum event or particle. And an electron could be the cause, causing the vial to crack if it has up spin and not if it has down spin. The electron can then be prepared in a superposition of states, under which a calculation can result in both up and down spin. However, if the spin isn't known before the measurement, then the state of the cat is also unknown. There's no way to tell whether it's living or dead in any realistic way. And that is clearly absurd.

Schrödinger's argument wasn't just that when quantum laws are enforced on a small scale, they seem to be nonsense — you do not need a pet for that. Instead, he tried to find a drastic example of how delaying the allocation of a definite condition (dead or alive) before calculation (by lifting the lid to see) might lead to consequences that are not just strange but also theoretically prohibited.

Chapter 5- The Double slit Experiment

The double-slit procedure is among the most well-known physics experiments. It shows, with unrivaled strangeness, that small particles have a wave-like quality to them, and indicates that simply looking at a particle has a significant impact on its nature.

5.1 The Experiment

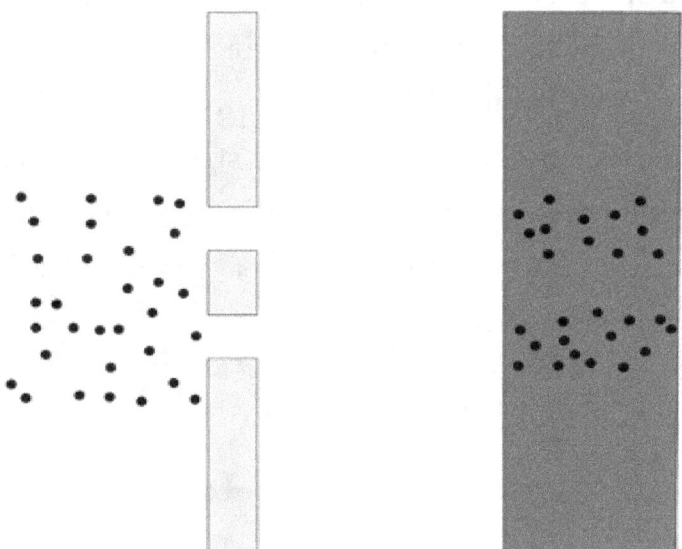

Envision a wall having two slits in there to begin. Consider hitting the wall with tennis balls. Some will hit the wall, and others will pass via the slits. The balls which have passed thru the slits can strike another wall if it has one behind the first. What should you hope to see if you tag all the places that a ball has collided with the second wall? Two parallel strips of tags of the same design as the slits.

Consider focusing a light (of a particular color, and wavelength) at a two-slit wall (in which the space between the slits is approximately equal to light's wavelength). The wall and light wave are seen from above in the picture below. The wave peaks are shown by the blue lines. The wave breaks into two waves as it goes through each slit, each extending out from the slit. The waves then collide with one another and cause interference. Where a peak approaches a trough, they cancel each other. For other times, when peak hits peak (as seen in the diagram by the blue curves crossing), they will enhance one another. The strongest light comes from areas where the waves reinforce one another. As light strikes a second wall placed behind the initial, it creates a striped pattern known as an interference pattern. The bright streaks are caused by the waves' mutual reinforcement.

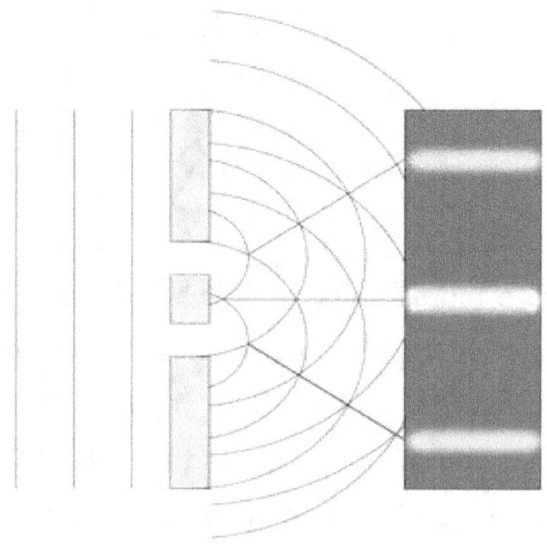

Below is an illustration of a real-life intrusion pattern. Since the photo shows more details than our description, there are many more stripes. (For completeness, we should mention that the picture also reveals a diffraction pattern, which will be generated by a single slit, but we won't go through it here, and you don't have to worry about it.)

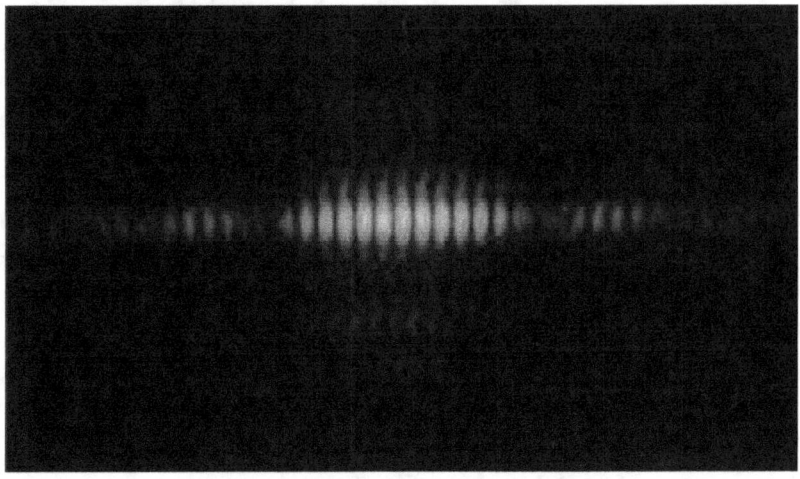

Let us now enter the quantum world. Imagine shooting electrons at our wall across the two slits, but obstructing one of them for the time being. Any electrons will travel through the open slit and touch the second wall in the same way as tennis balls do: the spots they land on create a strip that is roughly the same shape as the slit.

Open the second slit now. As for the tennis balls, you'd imagine two rectangular stripes on the second wall, but what you really see is very different: the places that electrons strike build-up to mimic the interference pattern of a wave.

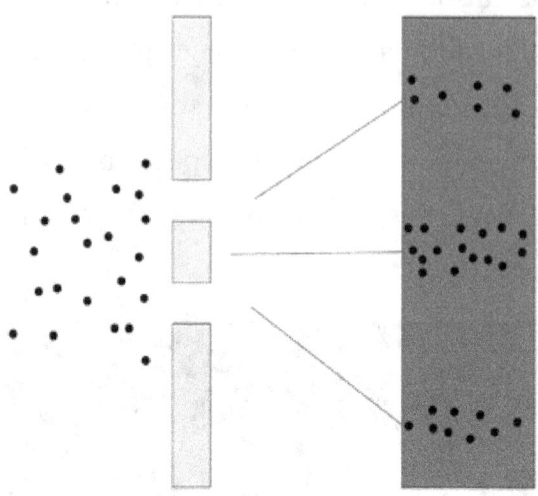

Here's a picture of a true electron double-slit experiment. Individual images depict the pattern that emerges on the second wall as more electrons are fired. As a consequence, a striped interference pattern emerges.

Let us now enter the quantum world. Imagine shooting electrons at a wall with two slits, but close one slit for the time being. Some electrons would travel via the open slit and touch the second wall in the same way as tennis balls do: the spots they land on create a strip that is approximately the very same design as the slit.

Unblock the second slit now. Just like the tennis balls, you'd imagine two rectangular stripes on the second wall, but what you really see is quite different: the places that electrons strike build-up to mimic the interference pattern of a wave.

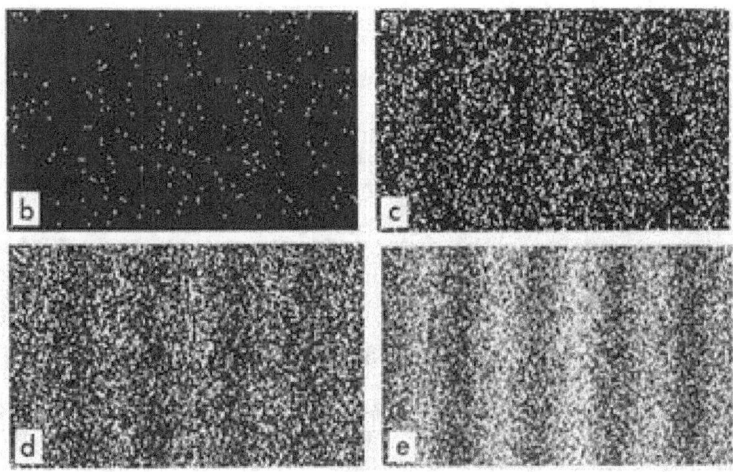

Above is a picture of a true double-slit electron experiment. Specific images depict the pattern that emerges on the back wall because more electrons are shot. As a consequence, a striped interference pattern emerges.

How is that possible?

One explanation is that the electrons interact with each other, causing them to arrive at different locations than they'd have they were isolated. The interference pattern, though, persists even though the electrons are fired one by one, with no risk of interfering. Surprisingly, each independent electron adds one dot to a larger pattern that resembles a wave's interference pattern.

Is it possible that each electron separates, passes into both slits at the same time, interferes with itself, and then reassembles to reach the second wall as an individual, localized particle?

You may use a detector near the slits and detect which slit the electron travels through. And that's when it gets strange. If you're doing that, the pattern on the screen of the detector will transform into a two-strip particle pattern, as shown in the above first picture! The interruption pattern is no longer visible. The process of observing somehow ensures that the electron particles act like tennis balls. It is as though they were aware that they were being watched and chose to avoid being captured in the act of doing strange quantum antics.

What do we learn from the experiment? It means that what we term "particles," like electrons, mix particle and wave properties in some way. This is quantum mechanics' well-known wave-particle duality. It also implies that studying and measuring a quantum field has a significant impact on the system. The measuring dilemma in quantum mechanics is the issue of just how this occurs.

The double-slit quantum experiment, as per eminent scientist Richard Feynman, brings humanity "up against paradoxes, anomalies, and peculiarities of existence." According to Feynman's rationale, if we could figure out what's occurring in this misleadingly simplistic experiment, we'd be able to get to the core of quantum mechanics — and even solve all of its mysteries.

The original experiment was classical, undertaken by British Thomas Young in the 1800s to demonstrate that light is a wave.

He shone light through two similarly distanced parallel slits on a screen, revealing many vivid bands on the other side. It was an 'interference' trend, he noticed. It's similar to the pattern that occurs as two rocks are tossed into a river and the ripples they produce contribute to or soften each other's troughs and peaks. The slits will behave more like stencils for spray paint with ordinary particles, forming two distinct bands.

Quantum particles, too, produce such an interference pattern, indicating that they also have a wave-like origin. This theory was proposed by Frenchman Louis de Broglie around 1924, and it was confirmed for electrons by Lester Germer and Clinton Davisson a few years later. This is true also for big molecules like buckminsterfullerene, which has 60 carbon atoms. And if particles pass through the slits one at a time, the interaction trend persists, accumulating over several particle impacts. The particles appear to collide with one another. Stranger still, if we use a detector to determine the slit the particle passes into, the pattern vanishes: the particle becomes completely particle-y, with no wave pattern. Surprisingly, this is true even though the calculation is delayed until after the projectile has passed through the slits (but before hitting the screen). Interference returns if we take the test but then discard the output without examining it.

It seems that it is the "effect of noticing," as scientist Weizsäcker (who collaborated extensively with quantum visionary Werner Heisenberg) called it in 1941, that makes the difference. This is what makes quantum mechanics so peculiar: it seems difficult to eradicate a definitive position for our conscious interference in the result of experiments. This fact led Eugene Wigner to speculate that the mind is responsible for the 'collapse' that transforms waves into particles.

Bohr's response was that quantum physics forbids one from predicting the particle's 'direction' — one or two slits — until it is calculated. According to Bohr, the theory's job is to include forecasts of measurement results, and it's never struggled in this regard. (Even so, he didn't dispute that there's any actual truth outside measuring, as is often assumed.) However, this seems to be unsatisfactory. David Bohm's alternate theory from the 1950s says Quantum phenomena are both wave and particle in this case, with the wave 'piloting' the particle across space thus becoming vulnerable to forces outside of the particle's physical position. Physics is yet to finish its journey through the double-slitted experiment. The mystery is yet to be solved."

With apologies to those researchers who believe they know the solution, this is correct: there isn't any consensus. In either case, Bohr was right in cautioning us about how we use words.

There is little in quantum mechanics that allows one to talk about particles being waves or following two routes at the same time as it stands without interpretation. There's no need to think about the wave function as anything more than an approximation. Roland Omnès characterized this function, which encapsulates all we understand about such a quantum particle (and appears in the classic equation invented by Schrödinger to explain the particle's wave behavior). It's "the gasoline of a device that creates probabilities," he says, referring to the probabilistic calculation outcomes.

5.2 Do Atoms Passing Across A Double-Slit Know They're Being Watched?

In a double-slit experiment, does a large quantum particle – like an atom – respond differently based on where it is observed? This question was posed in 1978 by John Wheeler's famed "delayed choice" by Gedanken experiment, and the solution has been recognized for the first time with large particles experimentally. The outcome shows that it's pointless to determine if a large particle's particle or wave behavior can be represented before a measurement has been taken. The methods used may be useful in potential physics studies, as well as possibly in knowledge theory.

Single objects, like photons, travel across a screen with two slits one at a time in the popular experiment. A photon seems to travel through one slit or the other if either direction is tracked, and no interruption is visible. If none of these slits are tested, a photon would seem to have gone through both slits at the same time before interacting with itself and behaving as a ripple. John Wheeler suggested a sequence of hypothetical experiments in 1978, in which he asked if a particle passing through a slit might be assumed to have a well-defined direction, in which it goes through either or both slits. The decision to track the photons is taken only after they've been released in the tests, allowing the observer's results to be tested.

What occurs, for example, if the particle commits to passing through one or both slits before the move to close or open one of them is made? If an interruption pattern remains after the second slit is opened, we must either assume that our action to measure the particle's direction has an impact on its previous decision as to which direction to follow or discard the classical principle that a particle's location is determined independently of our calculation.

Photon first

Although Wheeler thought of this as a thinking experiment, Alain Aspect was able to conduct it with individual photons in 2007, utilizing beam splitters instead of the slits Wheeler proposed. The researchers can recombine or isolate the two pathways by arbitrarily adding or withdrawing a second beam splitter, rendering it unlikely for an observer to determine which direction a photon has taken. They demonstrated that an interference pattern could be produced by inserting a second beam splitter after the photon had gone through the first.

Quantum mechanics' wave-particle duality states that all quantum things, large or not, will act either as particles or waves. Instead of photons deflected by beam splitters and mirrors, Andrew Truscott at Australian National University replicated Wheeler's work utilizing atoms deflected via laser pulses. The helium atoms were freed one at a time from an optical dipole trap and dropped under gravity till they were deflected through an identical superposition of 2 momentum states moving in separate paths with an adjustable phase gap by a laser pulse. The first "beam splitter" was this. Using a random-number quantum generator, the scientists then determine whether to introduce another laser pulse to merge the two states and establish mixed states – one created by introducing the two waves and the other created by subtracting them. When this final laser pulse was used, it was difficult to know which of the two directions the photon had taken when it was applied. The team replicated the experiment, changing the phase difference between both paths each time.

Double pulse

When the secondary laser pulse wasn't used, Truscott's team discovered that the chance of the atom being observed in any of the momentum states was 1/2, irrespective of the phase gap between them. The second pulse, on the other hand, created a distinct interference sine-wave pattern. When the waves arrived at the beam splitter precisely in phase, they intervened constructively, only maintaining the state created by adding them. Whenever the waves had been in antiphase, though, they interacted destructively, and the condition created by subtracting them was still identified. This suggests that embracing our traditional intuition of particles following well-defined pathways will compel us to consider backward causation. "I can't prove that isn't what happens," Truscott says, "and yet 99.999 percent of physicists would say that the calculation – i.e. if the beam splitter is out or in – takes the observable into existence, and the object determines if to be a particle or wave at that point."

Evidently, both Aspect's and Truscott's observations prove that an object's particle or wave nature is most definitely undefinable before a measurement is taken. Backward causation – that the photon has knowledge from the future – is a less probable possibility, as it requires transmitting a signal faster than the speed of light, which is prohibited by the laws of relativity.

Aspect is enthralled. "It's very good work," he notes. "Definitely, there is no great surprise about this kind of thing," he adds, "but it's a wonderful achievement." He goes on to say that, apart from being fascinating, the technologies created might have realistic applications. "Being able to master single atoms with certain precision can be valuable in quantum theory," he states.

At this point, I hope that everything you have read is clear to you and that you can leave a positive review that will allow me to get to know my creation to more people.

Chapter 6- Development of the Quantum Concept

Any of the theories we've looked at so far demonstrates how classical theories fall apart as we venture into areas beyond popular experience. When there are systems that fly very quickly, or when we travel through very heavy gravity, or when we imagine cosmic swaths of space, traditional Newtonian physics struggles. In realms of extremely high speeds, special relativity reigns supreme; in territories of very heavy gravitation, general relativity reigns supreme; and relativistic cosmology reigns supreme over vast distances.

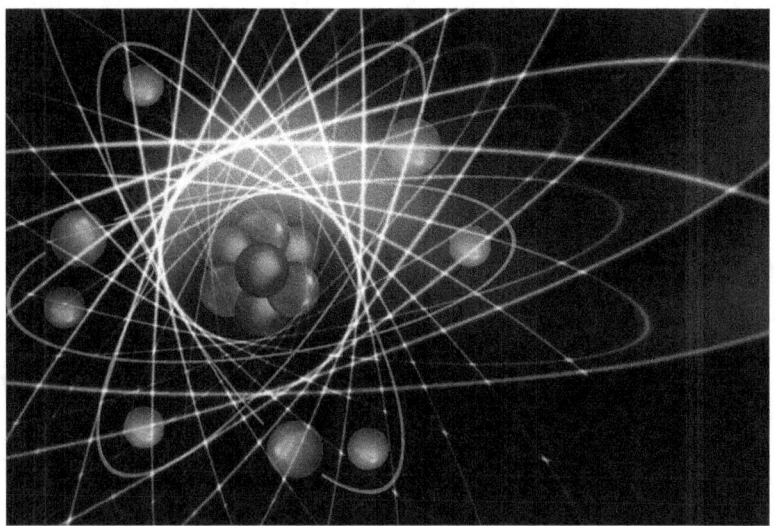

When we observe very tiny structures, such as actual particles and the atoms that make them up, traditional Newtonian physics collapses. Quantum theory offers us the most accurate picture of existence at the atomic level. The basic quantum theory we'll look at here allows no improvements to relativity theory's theories about time and space. The majority of standard quantum concepts are expressed in spaces and periods that are consistent with the special theory of relativity. Although several variants of quantum theory are embedded in general relativity space-time, full integration of quantum mechanics and Einstein's general relativity is still a long way off.

Quantum theory is a matter theory, or more specifically, a theory of the tiny components that make up everyday matter. Tables and benches, baked potatoes, and cows are also made up of objects such as electrons, neutrons, and protons, Quantum theory gives us the most accurate picture of these particles. It also gives us information about matter that is radiated, such as light. It is well understood that light is made up of both wave and particle. Quantum theory gave rise to the concept of photons.

The interpretation of the state of particles is the fundamental mystery of quantum theory. This state, it points out, doesn't really perfectly correspond to every state we are acquainted with from traditional physics. The objects in quantum theory resemble small tiny bits of matter in certain respects, as the word "particle" means. They resemble small packets of waves in some. To complete the picture, we must recognize that elementary particles possess both behaviors at the same time.

There is no simple way to imagine this required combination; in fact, there might be no picture that is entirely acceptable at all. The issue of getting there is always a problem today. The challenge, on the other hand, has proven to be no impediment to the theory's validity. Modern quantum mechanics has had a lot of scientific progress, explaining a lot of things and producing some interesting predictions.

The fundamental posits of quantum theory can be concisely stated. These assertions, on the other hand, are likely to seem arbitrary and often perplexing at first. What's required is a grasp of why such posits were selected and what issues they're supposed to fix. Reviewing the historical advances that contributed to quantum theory during the first years of the 20th century is the only approach to arrive at this interpretation. Since there is a naturally increasing series of challenges and solutions in the evolution that ultimately issues in modern philosophy. It's worth considering that this historical method is still preferred, but it wasn't included in this presentation of general and relativity. The reason for this is that the fundamental phenomena about which such hypotheses refer are either well-known or need no explanation. We can all see objects going very quickly, for instance, and it's just a small leap to believe that we won't be able to move these very rapidly moving things to light speed.

In quantum theory, Matters are different. The forces that govern the theory are mostly unknown. Outside of science, few people understand Ehrenfest's "catastrophe in ultraviolet" of heat radiation, or the strange frequency dependence of the photoelectric effect, or why the emission spectrum line's discreteness is traditionally worrisome. The historical solution acquaints us with these long-standing puzzles. When the resolution manifests as quantum theory, the nature and function of the resolution become readily obvious.

Unlike the theory of relativity, the quantum theory took a long time to develop and involved several hands. 20t Many scientists, like Einstein, contributed to its creation during the initial quarter of the 20th century.

6.1 Matter Theories at the End of the 19th Century

In the late 19th century, matter was recognized to come in 2 types.

The first; particles, which were little lumps of matter that floated about like small bullets. The electron was the most studied of the basic particles. In 1896, Thomson discovered that the cathode rays used in CRTs—the predecessor of glass TV tubes—were diverted by electromagnetic fields just like tiny small chunks of matter which is electrically charged. Atoms, which were made up of a clump of different particles, were thought to be particulate in nature.

The other kind of matter was wavelike. Light, or more broadly, electromagnetic waves, was the one well-studied type. Light, according to Newton and several other scientists in the 17th century, is made up of a storm of little corpuscles. Despite the fact that the wave perspective was also being followed at the time, Newton's corpuscular perspective remained influential. With the discovery of interference by Thomas Young at the turn of the 19th century, that changed.

The two-slit experiment produces the most well-known interruption effect. Light waves (shown as parallel wave fronts traveling up the screen) collide with a barrier that has two holes in it. Secondary waves emit out of the slits and overlap with one another, creating a distinctive cross-hatching design. The ripples cast out by two marbles dropping in the water create the same shapes like those found on the top of a still pond. The way the waves interact is crucial in these interference studies. The trends emerge as a result of the waves' ability to add up in two different forms.

The phases of the waves converge to produce a cumulative wave of higher amplitude in positive interference.

The periods of negative interference will be such that waves deduct and balance each other out.

Both disruptive and positive interference occur in various areas of the area where the waves converge in normal scenarios of interference, like the two-slit tests. As a result, the intricate interference structures can be seen emerging.

If you think about a wave as a displacement in a system, interference results are easy to consider. For e.g., a water tide in the sea has troughs and peaks where the water is moved below and above the mean sea level. When two waves collide and their tops coincide, the effect is a peak that is the sum of their heights. It is a good example of positive interference. When a rise and trough occur at the same time, the two will balance each other out. It is what is known as a disruptive intrusion.

Maxwell felt the theory of intervention was so convincing in the 19th century that it offered strong justification for an ether. If light is to have troughs and peaks that will balance out, he argued, there has to be a displacement in something of light. The ether is something that carries the light ray. It seems difficult to merge two corpuscles and make them cancel if light was composed of corpuscles.

With the ether theory's downfall, it became apparent that something more exciting was on the horizon. Light itself was created in such a way that it might locally cancel out other light waves. This type of interaction foreshadowed the types of interactions that would become prevalent in quantum mechanics.

This cool distinction between wave-like and particle-like matter does not last. The tale of quantum theory's arrival is the story of this division's disintegration. We'll see how numerous hints in observed physical characteristics of matter revealed that this simplistic division would fail in the sections ahead.

6.2 The Issue of Black Body Radiation and Max Planck

Radiation of Heat

In 1900, the first hint that radiation could have particle-like properties appeared. It was discovered during seemingly harmless research on heat radiation. Everybody is familiar with this kind of radiation. It is radiation which warms our palms next to a torch, cooks the toast, and gives a fireplace the intense flare. Physicists have been calculating the amount of energy contained in any of the various wavelengths (or colors) that make up heat radiation. The intensity of the radiation affects the distribution. As an object that generates radiation changes from white to orange to red heat, the frequencies with energy change accordingly.

The amount of energy that came from different areas of the color spectrum could be constructed by plotting. One of the graphs created by Otto Lummer and Ernst Pringsheim, two of the most prominent experimenters at the period. The color of the radiation is determined by the major peak in the gradient, which correlates to the spot in the continuum of much of the energy.

Pringsheim and Lummer are demonstrating that the evidence they discovered is not well matched by the structures of the strongest existing theories.

Max Planck was focusing on discovering the physical mechanisms that contributed to these energy distributions in 1900 when the latest and more recent data arrived. He was very well informed of Pringsheim and Lummer's most recent findings, and that no latest theory matched their most recent experimental evidence. He came up with a new application that worked well. Heat radiation, according to him, is a jumble of several wavelengths of electromagnetic waves which have converged in a cavity. Forces in the cavity's walls are absorbed and released by oscillating charges. The heat intensity of the walls will therefore be communicated to the radiation. The cavity is essentially an oven that radiates heat into the internal vacuum. "Cavity radiation" was the name given to the radiation generated within the cavity.

If a little window in the cavity's walls was raised, the radiation emitted will also have the cavity's temperature. It has the very same structure as radiation re-emitted by an object at the same temperature if the body had the unique ability to absorb completely all radiation that falls on itself before re-radiating it, according to certain creative thermodynamic claims. Since certain bodies are referred to as "black," the type of radiation is referred to as "black body radiation."

Planck's 1900 Study

Planck devised a straightforward formula that closely matched the most recent scientific findings. His issue was telling a hypothetical tale of how the formula came to be. He eventually came across such a tale after some delay. However, the central calculation in his narrative was based on an unusual premise. (Whether Planck knew how revolutionary this hypothesis was and how important it was to his accounts is still debated today.) Heat radiation was modeled by Planck as originating from energized electric resonators.

In classical physics, ordinary resonators are simply masses vibrating on springs, as seen in the diagram. They have the ability to absorb a wide variety of energies.

According to Planck's tale, these resonators could not be energized over a wide spectrum of energies. Instead, they could use energies ranging from 0 to 1, 2, 3, and so forth, but none in between. Energy levels of 3.2 or 2.7 units were forbidden.

It was crucial to figure out what those units were. The resonator's resonant frequency was used to determine the energy units. Planck's rule provided them:

h x frequency = energy

The approved energies are (frequency x h), twice (frequency x h), three times (frequency x h), and so forth.

The h stands for "Planck's constant," a modern natural constant invented by Planck and now known as such. This latest constant serves the same purpose in quantum theory as the light speed does in relativity theory: it indicates when quantum effects are likely to be important. The quantity is very tiny, implying that quantum effects ought to be found in the tiny; for instance, the units of energy provided by Planck's formula would be very tiny for normal frequencies, so we won't consider the granularity it necessitates as we look at the greater energies of systems encountered in everyday life. (h = 6.62×10^{27})

The energy of resonators was calculated using Planck's initial formula. He made every effort to limit the discontinuity it implied to these resonators, and also only the relationship between resonators and radiation. Other physicists realized during the next decades that the discontinuity cannot be contained. Heat radiation could be explicitly computed using methods similar to those used by Planck in 1900. They came to the realization that Planck's formula could also be extended to heat radiation. Heat radiation energy must be measured for whole units of frequency x h at each frequency. It's difficult to justify this conclusion with the assumption that radiation of heat is solely a wave process.

6.3 Light Quantum and Albert Einstein

The Proposition

Although Planck could not have known how revolutionary his 1900 study was, Einstein knew that something strange was going on with high-frequency light, and he did so seemingly independently of Planck. In 1905, he proposed that we ought to rethink our fundamental understanding of how radiation works.

In some cases, high-frequency light appears as if it is composed of spatially localized bundles of energy, with Planck's theorem used to calculate the sum of energy for each bundle. This was a shocking 180-degree switch. At first sight, it seems like we have been transported backward in history to Newton's old hypothesis that light is made up of a storm of corpuscles.

Einstein said that the energy of every corpuscle is equivalent to (light frequency x h).

High-frequency light, Einstein now argued, often acted as though it were composed of spatially localized bundles of energy. The sum of energy in each package was calculated using Planck's formula. As a result, light was once again defined as a rain of corpuscles.

Einstein's theory included a wave-based notion of frequency. Or whatever else may arrive, the studies on the light interference remained. Einstein's central point was brilliant. As he examined the characteristics of high-frequency light, he found that they were regulated in certain ways by the same rules that regulate ordinary gases. By reversing the gas theories, Einstein was able to prove that they were based on gases made up of a large number of spatially localized tiny lumps of matter called molecules. He believed it was no coincidence that gases and light followed the same rules; they did, he argued, since light was composed up of tiny, localized units of energy known as "quanta."

The term "quantum" (plural is "quanta") was therefore simply applied to a unit of measurement. Speaking of a "light unit" in 1905 was understood to be the same as talk of a "light quantum."

The Photoelectric Effect

Einstein's 1905 article on the light quantum is better remembered for an assertion made near the end of the article. Einstein had been interested in research involving the "photoelectric effect." Light is used to expel electrons from an electrical cathode in this experiment. The strength of light, as per the wave theory of light, can decide if it will produce these "photoelectrons." Since more concentrated light has greater energy, it is needed to free the electrons trapped on the cathode's layer.

It is simple to reduce the intensity of light. For e.g., we may simply transfer the light source further away such that the energy it releases is distributed across a large region. The wave hypothesis predicts that the light would reduce its capacity to liberate photoelectrons when it dims.

However, experiments have shown that the intensity of light had little bearing on its capacity to emit photoelectrons. All that counted was the light's frequency. And if the light was really bright, it could not produce photoelectrons if it had a low frequency. Even though the light was of really low intensity, it might emit photoelectrons when it has a high enough frequency.

This is exactly what one would predict if light energy were localized in quanta of energy provided by Planck's formula, as Einstein proudly observed. To produce each photoelectron, all one had to do was presume that only a singular quanta was needed.

If the light has a low frequency, each quanta has low energy, but no single quanta is powerful enough to push electrons out from the cathode. Raising the light intensity simply increased the amount of light quantas hitting the cathode, both of which were too low in energy to free a photoelectron.

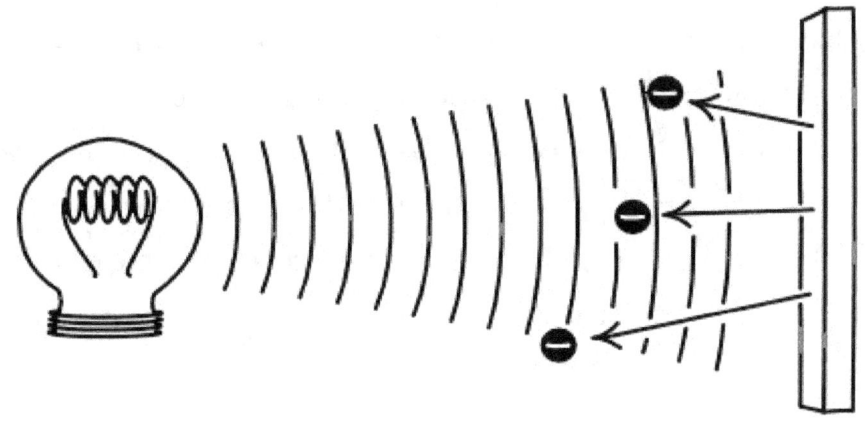

Each light quantum was independently powerful enough to unleash a single photoelectron if the lighting was of higher frequencies. It didn't matter how bright the light was. There were few light quanta hitting on the cathode due to the low intensity. However, because only a single light quantum is required to free one photoelectron, the effect will be present for high-frequency light, regardless of its intensity.

Einstein was awarded the Nobel Prize in physics, in 1921, 15 years later. The award specifically mentioned his study on the photoelectric effect. "For his contributions to Physics, notably for his realization of the rule of the photoelectric effect," the citations read.

The light quantum theory proposed by Einstein was bold. It was even more audacious than that.

His other 1905 research contributed to the completion of 19th-century physics.

His Brownian motion report and doctoral dissertation were instrumental in establishing the reality atoms. This accelerated the mathematical physics study of Boltzmann, Maxwell, and more in the 19th century. They wanted to prove that statistical study of the action of a large number of constituent molecules or other elements could be used to recover the measurable thermal state of materials.

Modern kinematics of space and time was represented inside the electrodynamics of Lorentz and Maxwell. It was derived and displayed from Einstein's thesis on special relativity.

However, Einstein's proposed light quantum posed a strong threat to the Lorentz-Maxwell tradition's crowning achievement: the finding that light was simply a wave propagating through the electromagnetic field.

Was Einstein aware of his light quantum hypothesis's unique audacity?

These types of queries are rarely simple to address. In this situation, though, we have ample proof that Einstein was aware of the document's audacity. Einstein wrote an unforgettable note to his colleague Conrad Habicht in 1905. It evaluated a fascinating selection of papers he was publishing.

He simply states that the kinematic section of the paper "will undoubtedly pique your interest." He tells Habicht, however, that the light quantum report is "really groundbreaking."

6.4 Duality in Waves and Particles

Do we really need the wave view if this corpuscular perception of light is so efficient? Einstein demonstrated in 1909 that such effects could only be described successfully when both the particle and wave views were combined; the total observable influence originated from the sum of two things, one a wave term and another a particle term. The requirement for both is often referred to as "wave-particle duality," and anything like this is needed since, and whatever our reports of light are doing, they must also satisfy Thomas Young's findings of interference anomalies of light from the early nineteenth century.

Many of you would like to use the term "photon" synonymously with Einstein's "light quantum," which is well as long as you understand that the term "photon" dates from a later period of quantum theory. G Lewis was the one that brought it up in 1926, after a dramatic 21 years.

The natural assumption is that when we say photon, we're talking to the object that emerged from the concluded quantum theory of the 1930s. Even Einstein himself may not have seen how far quantum theory will diverge from traditional concepts when he suggested his light quanta. His 1905 theory was very limited; he proposed that the energy of light of high frequency was spatially condensed into small lumps known as light quanta. He had no idea how it would turn out for light of low frequency. And his 1905 proposal made no mention of momentum of light quanta. Later, it was discovered that light quanta had momentum as well.

6.5 Atomic Spectra and Niels Bohr

Heat radiation detection and the ability of light to produce photoelectrons offered the first indications that this wavelike sort of matter was not only wavelike but it's also particle-like. What about matter's constituent particles? What about the electrons discovered by Thomson in 1896? Findings in atomic spectra provided the first hint that they had wavelike properties.

As gases are heated or have an electric discharge passed into them, they will emit light. This process is found in the simplest terms in parking lot orange sodium vapor lamps and clear white mercury vapor lamps. The mechanism can also be reversed. Gases consume light, which is how they may obstruct light transmission.

6.6 Matter Waves of Schrodinger and de Broglie

Image: Erwin Schrodinger

Waves of Matter

Bohr's 1913 theory and subsequent extensions provided a fantastically diverse collection of techniques for atomic spectra accounting. They were based on a jumble of classical and non-classical ideas. By the 1920s, the system's limitations had become apparent, and researchers were tasked with making analysis of this theory, which became recognized as "old quantum theory."

The "new quantum theory" breakthroughs came in the mid-1920s. A lot of different researchers discovered ways to build consistent quantum domain theories, and they all turned out to be different variants of the same new hypothesis in the end. Matrix mechanics was invented by Heisenberg, Jordan, and Born. Its fundamental quantities were endless tables of numbers called matrices, which were derived as accurately as possible by observed quantities as atomic spectra.

Another method was found to be similar and is simpler to visualize. It was established by Schrodinger in 1926, based on a hypothesis by de Broglie in 1923. Einstein demonstrated that light, a wave phenomenon, have particle-like properties. Is it possible that the opposite is also true? Is it possible that particles like protons have wave characteristics as well?

Yes, according to the theory. It linked a wave of a specific wavelength to a particle with a specific momentum.

De Broglie's method for determining which wavelength corresponds to which momentum is as follows:

H / wavelength = momentum

It's worth noting how close it is to Planck's equations for the relationship between frequency and energy. Here's Planck's formula once more:

h x frequency = energy

The matter-wave method is built on the basis of these two concepts. They explain how to match a particle's momentum and energy to a wave with a certain wavelength and frequency.

Here's another way to show the two calculations side by side. We may write 1/period = frequency for a periodic process, where "period" is the time required for the operation to repeat. Then Planck's formula takes on a new meaning.

H / period = energy

The equations now connect momentum to duration (the wavelength) and time interval (the period) to energy.

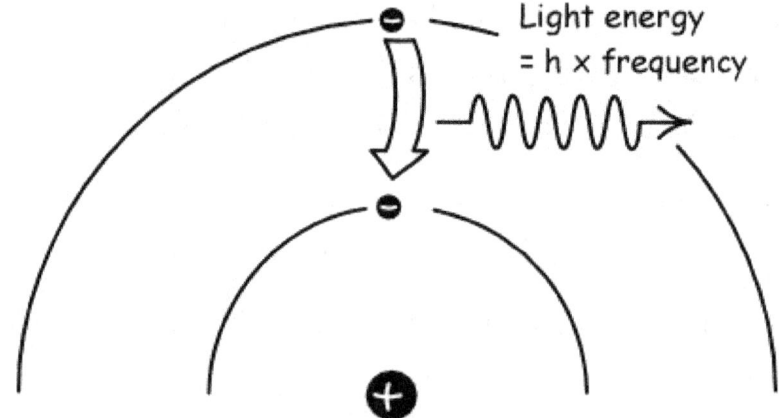

Atomic Electron Energy Discreteness

The uniqueness of the matter wave theory is that it explains why electrons embedded in atoms may only exist in very specific energy states. The fact that these electrons can only exist in some energy states stems directly from the basic distinctions between waves and particles. We could see these distinctions by taking into account a very typical example, a wave/particle stuck in a container.

To keep it easy, we'll assume that the particle could only travel in one dimension and that it is trapped within a 1-D box.

Consider a regular, classical particle enclosed in a box. Here between walls, it swings back backward and forwards. It can travel at any pace because of classical mechanics. As a consequence, it may have a wide spectrum of various energies at any given time.

Consider instead subjecting a wave to the very same container. "Standing waves" are the stationary waves that will remain inside the box.

They are familiar to anybody who performs a stringed instrument. As a string is pulled or bent, a standing wave with a half-wavelength equal to the string's duration produces the base tone. Overtones are often created, which add to the sound's depth. There are shorter standing waves, with wavelengths comparable to the string's duration, 2/3 of the string's length, a fifth of the string's length, and so on. A wave will shape as long as there are nodes—points with no displacement—at either end of the string.

Such tones and overtones have the same shape as the waves of matter that may develop inside the container. On they contain wavelengths of one, half, one-third, and so on, multiplied by the container's double width. Each one of these waves has separate energy that is dependent on the standing wave's wavelength. As a result, only a few definite energies for the waves stuck in the container are allowed; the several intermediary energies between them aren't

What about de Broglie's formula, h/wavelength=momentum? Is it correct to assume that the momenta of the standing waves in the box are corresponding to h, h/2, 2h, (3/2) h,... etc., referring to the permitted wavelengths mentioned above. At least, that's how it feels. form A particle going to the left with h/(width of the box)=momentum and the other moving to the right with the same momentum formulas could be correlated with a standing wave with wavelength equivalent to the width of the box. A standing wave, on the other hand, would not propagate to the right or left.

We build a superposition of these couple waves to make the wave remain still. We may get a wave that moves simultaneously to the right and left at the same moment, and hence goes nowhere, thanks to superposition.

The scenario is exactly the same with an electron inside a hydrogen atom. The positively charged nucleus's electric attraction creates a cage for the electron, similar to how the box above captures the wave. Just a few energy states of the wave in the container may persist. As a result, an electron-wave stuck in such a hydrogen atom will only exist in a few distinct energy states. These happen to be energies of Bohr's theory's stable orbits.

Although such energies are preserved, Bohr's theory does not preserve the notion of the electron as a spatially localized particle circling the nucleus in an elliptical orbit or classical circular while breaking traditional electrodynamics by just not radiating. A standing stream of electrons fills the vacuum around the nucleus of the atom. The traditional electrodynamic theory, therefore, doesn't exist explicitly, and the earlier inconsistency with it has vanished.

6.7 New Quantum Theory

All of these theories came together in the late 1920s to form the "new quantum theory," which was distinguished from the "old quantum theory" of the previous decades. There were several

- Heisenberg, Jordan, and Born's matrix-based strategies; and
- Schrodinger and de Broglie and's matter waves; and
- Dirac's "c-numbers" (classical) and "q-numbers" (quantum).

It wasn't long until it became apparent that all of these concepts were actually just separate mathematical versions of the same principle. The enigmatic characteristics of matter and light that had contributed to this explanation had now been largely overcome. The answer rests in a new understanding of matter's existence. Matter is made up of a kind of matter which is both wave and particle, and this is so with both regular matter like electrons and protons & radiative matter such as light.

However, this recent synthesis left a residue of lingering issues.

First, the modern theory added a probability element that had never been used before in classical physics. Many systems have consequences about which the theory will only have probabilities. Is this radioactive atom going to decay right now or later? Probabilities are the most the hypothesis has to deliver. Many theorists of the day, like Einstein, were troubled by this situation. They considered it repulsive to consider that the universe's universal rules could be probabilistic, and they referred to the dilemma as a "causality collapse."

Second, for small particles, the modern quantum theory performed admirably. However, how it could be applicable to larger bodies was less obvious. Tables, beds, buildings, and elephants do not seem to have both particle and wave characteristics. They do, though, according to the theory.

Even if the strength of light around the continuum varies, one would assume those emissions (and absorptions) to include all wavelengths (colors)—a true rainbow. They don't. The frequencies that gases release and consume are extremely selective. They can only produce and ingest a few specific frequencies. The atomic emission range of the element is made up of the frequencies released, while the absorption continuum is made up of the frequencies consumed. They have distinct frequencies which may be utilized as a signature for distinguishing a gas that is otherwise unidentified.

The hydrogen gas emission spectrum is seen below. By moving the light released by hydrogen via a diffraction grating or prism, it's been separated into its frequencies. The light then blackens a photo emulsion in various locations based on its frequency. The lines shown are part of the "Balmer series," which occurs in visible light frequencies. (Wavelengths are measured in Angstroms.)

Rutherford's Nuclear Model Fails

Niels Bohr published a paper in 1913 about his attempts to develop a model of the method of light emission from elemental atoms that would describe the very specific frequencies released. The dilemma turned out to be much more difficult than expected. Rutherford's nuclear experiment was a great representation of the atom at the time. An atom, based on that information, is like a miniature solar system. It has a huge positive nucleus the nucleus attracts smaller, negative electrons that orbit it, similar to how planets orbit the huge sun.

As in the Rutherford model, energizing a gas by transmitting high volt electricity thru it energizes the electrons, allowing them to travel away from the nucleus's enticing pull. The energy gained will be released as light energy as they fall down back into the nucleus.

The first problem was that when they returned to the nucleus, they must travel across a continuous spectrum of orbital frequencies, emitting a continuous range of light frequencies. It was impossible to restrict the light released to only a few specific frequencies.

The second issue was even more extreme. Nothing will hinder the electrons from emitting energy through the method of light emission. They'd keep doing so until they collided with the nucleus. This can happen very easily, as per classical electrodynamics. Rutherford's paradigm did not seem to compensate for the existence of atom-based matter.

The Theory of Niels Bohr

With an idea of dazzling audacity, Bohr addressed both problems. Classical electrodynamics is straightforward: an electron circling the nucleus accelerates and would also emit radiation. It'd be similar to a small radio transmitter that broadcasts electromagnetic waves. This must waste energy in the process, sink further into the nucleus's enticing pull, and finally crash into the nucleus.

Bohr merely said that this wasn't the case. He stated, instead, there are stable orbits clustered around the atom's nucleus where an electron can orbit forever without losing energy.

Then, according to Bohr, electrons will jump between these permitted orbits. An electron must accumulate energy in order to hop away from the central nucleus and into a higher energy orbit. This allows it to transcend the atom's positive nucleus's attraction and travel away from all of this. Light dropping on the atom provides additional energy to it. The light's energy is passed to the electron, allowing it to leap to a high-energy orbit. The light should deliver precisely the correct quantity of energy to compensate for the energy gap between the orbitals.

Furthermore, Bohr believed that the volume of energy extracted from the thrilling light follows Planck's formula: energy equals h times frequency.

As a result, only light of a very particular frequency will excite the leaps between two orbitals. For a given leap, the light's frequency must be specifically calibrated such that frequency x h equals the energy required to achieve the leap. The reverse mechanism is also possible according to Bohr's hypothesis. An electron that has skipped to a higher orbit can't remain there indefinitely. It would descend to a lower orbital once more. It would then re-emit the energy it received from leaping up

as certain frequency of light. The energy of the light produced would be equivalent to h x frequency, as predicted by Planck's formula.

As a consequence, when an electron leaps between two orbitals, it absorbs light with a certain frequency that is exclusive to that leap.

These amounts of light that it consumes and releases eventually became known as light quanta (Einstein). Bohr, on the other hand, was careful in 1913. He made no mention of a light quantum being emitted. He merely stated

"The emitting of a relatively homogenous radiation where the Planck's theory gives the relationship between the amount of energy transmitted and frequency."

It wasn't by chance that Einstein's quantum was avoided. Bohr was a staunch critic of Einstein's theory and only changed his mind after modern advances in the 1920s forced Einstein's theory to be accepted.

Bohr was able to deduce the strangest finding from the measured atomic spectra after making certain assumptions. As only a few light frequencies were available, only a few hops were feasible, resulting in only a few orbitals being allowed for the electron. It was as if our sun permitted a comet to orbit between Mars and Earth, but barred every planet in between.

It was just a matter of determining which of the several alternative orbits are located in this favored set of stable orbitals. It was a pretty simple task. The differences of energy between these permitted, stable orbits were cataloged in detail by the observed spectra. Every line in the spectra was caused by electrons moving between two different orbits. It's a computational experiment to figure out which of the few orbits are which. The estimate was quite similar to this geography exercise. If we know the lengths between each couple of cities in a world, we can use that information to find out where each city is on the globe. The energy differences between Bohr's permitted orbits were determined by atomic spectra. He was able to calculate the energies and therefore the positions of the permitted orbits using these results.

When Bohr did this, he discovered a rather clear way to summarize the orbits. They were the ones whose angular momentum was expressed in h/2pi units. Bohr discovered that circling electrons always have to have whole units of angular momentum: two h/2pi, seven h/2pi, eleven h/2pi, with nothing in between, much like Planck's comparison tried to tell us that radiant energy is coming for whole units of frequency x h. We've shown that an object's ordinary momentum is equal to its mass multiplied by its velocity. Angular momentum is essential in the dynamics of spinning or orbital structures. It is known as an electron's orbital radius x angular speed x mass for a small particle like an electron circling a neutron.

Bohr's model was perplexing, to say the least. It appeared to demand that traditional physical concepts both retain and collapse at the same moment, much like Einstein's light quantum hypothesis. It was not a pleasant position to be in. Those annoyances were overshadowed by a more positive reality. The hypothesis of Niels Bohr succeeded, and it operated well enough. Observational spectroscopy provided researchers with a large database of spectra from a variety of samples under a variety of conditions. Scientists were able to construct an enormous amount and fruitful account of them based on Bohr's theory. Although it was obvious that something wasn't quite right, it was appealing to put off these worries in the face of excellent results.

The angular momentum that orbiting electrons possessed arrived in absolute multiples — quanta — in h/2pi, according to Bohr's theorem of 1913. In the years that followed, that simplistic condition was extended into a wider condition in which the quantity "action" was only available in entire multiples for physical processes that reverted to the same original condition on a periodic basis. As a consequence, physicists began to use the expression "quantum of action." This sidebar should have a single sentence that describes the physical quantity "action."

Chapter 7- Observing a Quantum System

One of quantum theory's most perplexing concepts, which has long intrigued theorists and physicists alike, says that the process of watching influences the observable truth.

7.1 Observation Affects Reality

Researchers at Weizmann Institute have now performed a highly structured experiment showing how the act of being watched affects a beam of electrons. The experiment showed that the more "watching" a person does, the more control he or she has over what really happens. Mordehai Heiblum led the study team.

Quantum mechanics says that particles will act as waves when observed by a quantum "observer." This is valid for electrons at the submicron scale, that is, at wavelengths of a thousandth of a millimeter. When acting like waves, they will travel through several holes in a barrier at the same time and then reassemble on the other hand. Interference is the term for this "meeting."

Interference will only happen while no one is looking, as odd as it might seem. As an observer continues to observe the particles pass through the holes, the image begins to transform dramatically: whenever a particle is observed passing through one window, it is apparent that it did not pass through another.

In other terms, as electrons are seen, they are "pressed" to act like particles rather than waves. As a result, the process of observing has an effect on the experimental results.

To show this, the Weizmann Institute team made a tiny system with two opens that were less than a micron in dimension. They sent over electron current towards the barrier. This experiment's "observer" was not human. For this, scientists at the Institute used a small but advanced electronic sensor that can track moving electrons. Changing the conductivity of the quantum "observer," or the frequency of the current flowing through it could change its ability to identify electrons.

The detector has little impact on current other than "detecting" or "observing" the electrons. However, the position of the detector-"observer" along one of the openings induced variations in the interference of electron waves traveling via the barrier's openings, according to the researchers. In reality, this impact was proportional to the "amount" of the observation: as the "observer's" potential to discern electrons improved, or as the degree of the observation improved, the interference reduced; conversely, as the "observer's" ability to discern electrons decreased, or as the observation softened, the interference enhanced.

Scientists were able to manipulate the magnitude of the quantum observer's effect on electron activity by manipulating its properties. A variety of scientists of the Weizmann Institute, as well as Tel Aviv University, formed the theoretical foundations for this phenomenon many years ago. The latest experimental project was performed after talks with Weizmann Institute, and the findings have already piqued the attention of theoretical physicists all over the world.

The technology of the Future

The discovery that observation continues to destroy interference can be utilized in future technologies to maintain data transmission confidentiality. This is possible if the information is stored in such a manner that decoding it requires the intervention of several electron routes.

On a bigger level, the Weizmann Institute research contributes significantly to the science community's attempts to build quantum electronic devices, which could become a possibility in the next century. This revolutionary new form of electronic hardware will take advantage of both the wave and particle natures of electrons at the same time, and a better understanding of how these two properties interact is required for its growth. For instance, future technologies could pave the way for the production of new computers with capacities well beyond those of today's technologically advanced machines.

7.2 What Does Quantum Theory Have to Say About Reality?

It was astonishingly simple for proof that reversed Isaac Newton's theories regarding the properties of light. The double-slit project, as defined by English scientist Thomas Young, "may be replicated with great ease, anywhere the sun shines," and Young was not being too dramatic. He'd devised a simple but beautiful experiment to demonstrate light's wavelike existence, refuting Newton's hypothesis that light is composed up of corpuscles.

However, the early 1900s saw the rise of quantum mechanics, which proved that it is made up of small, indivisible units of energy known as photons. When performed with single photons, Young's test is a fascinating puzzle, asking profound concerns regarding the origin of reality. Some have also used it to contend that human cognition influences the quantum realm, granting our minds power and a role in the universe's ontology. Is the basic experiment, however, sufficient to support this claim?

Young's experiment, in its current quantum shape, comprises beaming light or matter molecules at two slits carved through an otherwise impermeable wall. A display on the other side of the divide tracks the appearance of the particles. Photons must go through 1 slit or the other start piling up behind each slit, according to rational thinking.

They aren't. As we discussed multiple times, they gravitate to specific areas of the screen while excluding others, resulting in overlapping bands of light and shadow. There are interference fringes, which occur as different pairs of waves collide. Where the crests of two waves line up, you get positive interference (glowing bands), and then when the crests match up with troughs, we get negative interference (darkness).

For every given moment, though, just one photon passes into the device. It was as if each photon were passing through all slits at the same time, causing interference. This isn't consistent with classical logic.

However, what passes through both slits arithmetically is not a physical wave or particle, but a wave function; a math function that describes the status of a photon (in this scenario its location). The wave function acts in the same way as a wave. It collides with the two slits, causing fresh waves to emerge by each slit on the opposite side, which scatters and ultimately interact with one another. The probability of finding the photon can be calculated using the composite wave function.

The photon is more likely to be detected when the 2 wave functions interact constructively and is impossible to be located where they interfere destructively. The wave function is shown to "collapse" as a result of the measurement—in this situation, the engagement of the wave function with the screen. It shifts from being dispersed prior to analysis to peaking at one of the locations where the photon materializes.

Many mathematical problems in quantum theory stem from the obvious measurement-induced failure of the wave function. There's no way to know for sure when the photon would arrive before the collapse; it could occur in all of the locations with a non-zero likelihood. It's impossible to trace the photon's path between the detector and source.

The mathematics was perceived by Werner Heisenberg and others to suggest that truth does not occur unless it is witnessed. "It is difficult to imagine an actual natural universe whose smallest components function logically in the same way as mountains or plants exist, regardless if we see them," he wrote. John Wheeler used a double-slit experiment and argued that "no fundamental quantum phenomena is a phenomenon before it is a reported phenomenon."

However, quantum theory remains ambiguous on what defines a "measurement." It merely assumes that the measurement mechanism must be classical, without specifying where the distinction between quantum and classical resides, keeping it open for people who believe that human cognition must be asserted in order for the system to collapse. In this platform, Henry Stapp claimed that the double-slit test and contemporary versions show that "an aware observer may well be essential" to make any sense of the quantum domain and that a transpersonal consciousness resides underneath the material universe.

However, these experiments do not have scientific support for those assertions. Only the probabilistic assumptions of the mathematics can be checked in the double-slit test using single photons. The hypothesis argues that the wave function of each photon collapsed due to an ill-defined mechanism called estimation if the odds are confirmed over the course of passing tens of thousands of equivalent photons thru the slit. That is what there is to it.

There are many other interpretations of the double-slit. Take, for example, the de BroglieBohm principle, which asserts that reality is a particle and also a wave. A photon still has a fixed location when it approaches the double slit and passes through one of the slits; hence, every photon has a route. It's surfing a pilot wave that passes both slits, starts to interfere, and then directs the photon to a positive interference spot.

In 1979, in London, Chris Dewdney simulated the paths of particles passing the double slit as predicted by the theory. Experimenters have confirmed the existence of such paths in the last decade, although with the use of a problematic method known as weak measurements. About the debate, the experiments prove that the de BroglieBohm hypothesis is still in the running as a theory to explain quantum behavior.

Importantly, witnesses, measurements, or even non-material awareness are not needed for the theory to work.

Neither do collapse theories that claim that wave-functions collapse at random and that the larger the quantum systems number of particles, the more probable the collapse. Observers simply discover the end result. By passing bigger and bigger molecules through the slit, Markus Arndt in Austria has also been putting these hypotheses to the test. According to collapse theory, as particles grow huger than a certain threshold, they will no longer exist in a superposition of passing via both slits at the very same time, and the interference pattern would be destroyed. And after sending a molecule of over 700 atoms through the slit, Arndt's team sees interference. The pursuit for the dividing line continues.

Because of gravitational uncertainties, the larger the mass of a superpositioned object, the sooner it can fall to one condition or the other, according to Roger Penrose's iteration of a collapse concept. It's another observer-independent philosophy. There's no reason to be conscious. Dirk Bouwmeester is putting Penrose's theory to the test with a double-slit trial at the University of California.

The intention is to bring a photon in a superposition of passing through the slits at the same time, as well as one of them in a superposition of it being in 2 places at the same time. If the photon is in motion, the relocated slit will either remain in superposition or it may collapse, resulting in various forms of interference patterns, as per Penrose. The slit's mass will determine the collapse. Bouwmeester has spent a decade working on this experiment and will shortly be able to confirm or deny Penrose's arguments.

Even if the arguments are quite well philosophically or mathematically, these experiments show that we can't yet draw certain claims regarding the existence of reality. Considering that neuroscientists and philosophy of thought disagree on the essence of cognition, arguments that it collapses the wave functions are, at best, premature and, at worse, incorrect.

Chapter 8- Quantum Leaps

As discussed briefly before, Quantum jumps or leaps are energy level jumps made by a moving electron in an atom. As an electron travels to a lower energy state, the atom produces a photon; when it goes to a higher energy state or exits the atom, the atom receives a photon (ionization).

8.1 Reasons

In this segment, classical mechanics will describe two explanations for this quantized energy, all of which are linked to the spin of a proton. Protons and electrons have the same spin. There are 4 tetrahedral quarks along with an anti-quark in the updated proton pentaquark model. The tetrahedral quarks cancel spin (+ 1/2, + 1/2, − 1/2 − 1/2), allowing the positron/anti-quark in the middle to spin. It may have a measure of 1/2 or -1/2. The positron/anti-quark represents longitudinal waves that trigger the Coulomb force/electric force, however, its spin also produces another, transverse wave, as seen in red beneath.

In modern mechanics, spin has two directions, which are attributed to as spin-down and spin-up. The below icons are included in this explanation since quantum leaps are relevant to the configuration of protons inside the nucleus that is caused by their tetrahedral arrangement:

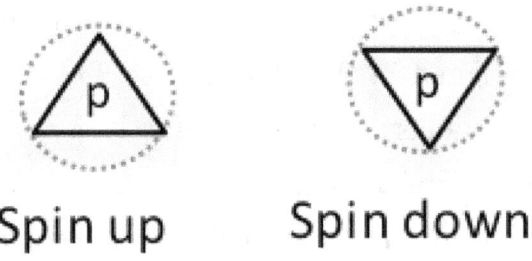

The spin-down and spin-up symbols for the proton indicate tetrahedral quark orientation.

Two or more than 2 spin-aligned protons

A wave traveling between two or more than two spin-aligned protons is the first trigger for quantum leaps. The orbital force increases, repelling the electron even further, relative to the square of protons in orientation.

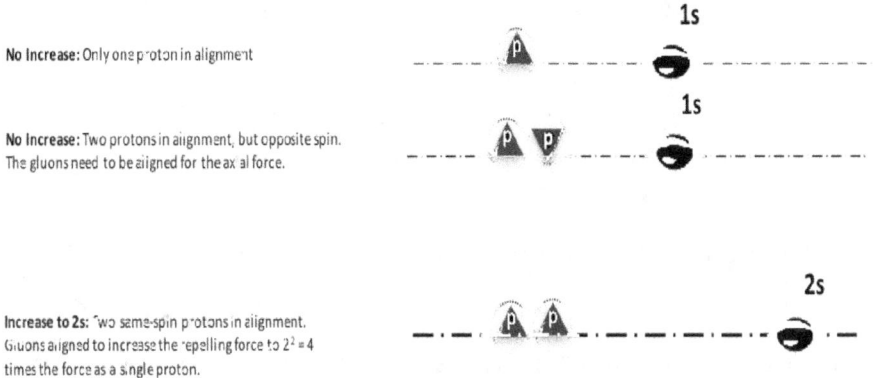

Three representations are shown above. The first is a wave that just passes across one proton. It creates the force repelling orbital towards the 1s orbital by passing between two quarks inside tetrahedral vertices. The tetrahedral vertices of two protons are aligned in the second case, but their tetrahedral vertices aren't thanks to opposing spins. As a result, it repels the 1s orbital. The orbital force is increased in the last case since two similar-spin protons have a wave that goes through an assembly of quarks. This pushes the electron into the 2s orbital, seen in the diagram below.

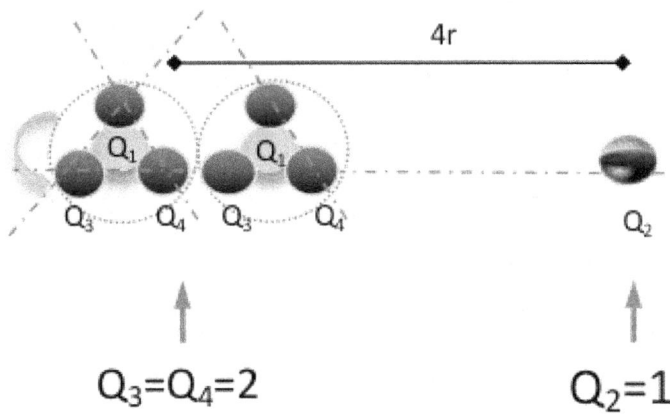

Protons with gluon in alignment is two. Q_3 and Q_4 are two. Q_2 is one (electron).

With the modified particle count Q4 and Q3, which reflects the number of protons in orientation, which is 2, the repelling, (F2) axial force are revisited now. Remember that Q2 is the count of particles for the single impacted electron. The orbital gap equations and measurements were based on these assumptions.

$$F_o = \frac{E_e r_e^2}{a_e^2}\left(\frac{(2)^2}{r^3}\right)$$

2 spin-aligned protons repel each other.

The composition and configuration of protons inside the nucleus of atoms determine the general source of multiple orbitals of different energies. In the ground state, hydrogen (having one proton) has the electron in the 1s orbital. Since there are 2 protons with opposing spins, helium (having two protons) also has its electron inside the 1s orbital. S2ince lithium (having three protons) has two similar-spin protons, one electron would be inside the 2s orbital in the axial way of these 2 protons.

Energy gain of a proton spin

An energy increase in the proton's spin can be a second source of the quantum jump. At ground state, hydrogen, for instance, has an electron in the 1s orbital (Bohr radius).

It influences the orbital power, which influences the orbital gap if it gains a photon (energy). The orbital force which repels is now 2 times greater with each tetrahedral quark in combination if the photons transverse wave induces the spin of proton to move at 2 revolutions every cycle (the longitudinal wavelength).

As a proton spins with 2 revolutions every wavelength, the hydrogen electron is energized to the 2s orbital.

Resonance is the cause of the quantum leaps in this situation. An increase in spin energy may allow an electron to shift orbitals or ionize, but as stated on the photons page, the spin power must resonate also with longitudinal wave power to keep rotating.

The frequency of the longitudinal wave synchronizes spin. It must be an integer of a periodic equation to be a resonance frequency. A carousel is an analogy. To successfully drive the pony at the right moment, the frequency of the longitudinal wave must balance the carousel's rotating frequency. Energy flows through and isn't transformed to spin energy if this is not the case.

Since the spin must follow the periodic longitudinal wave frequency, the amplitude of the longitudinal wave will increase and induce quicker spin, and it may only be in integers (1x, 4x, 7x). For example, wave amplitude of 2x equals rotational spin 2x; however, wave amplitude of 2.5 equals 2x rotational spin. As the electron settles back to the 1s orbital, it vibrates into place, producing a new photon with that same transverse force.

This resonance frequency and spin interpretation can now be mathematically modeled. It will now rotate at the frequency if the energy is adequate to exceed two revolutions every (2x) longitudinal wavelength. The equation is seen in the diagram below. The strong force's fine structure (αe) is the difference in amplitude of waves between quarks and has been changed to be 2x stronger to every quark in axial orientation. It's worth noting that the rise in wave amplitude, is the reciprocal of the constant of fine structure.

Despite the fact that the calculations used to calculate the force vary, the net effect will be the same force for both reasons. It's worth noting that Q4 and Q3 will still remain the same. The quantity of similar-spin protons aligned is denoted by the variable Q.

8.2 Controlling Quantum Leaps

What actually is quantum mechanics, and how does it work? Basically, it's the mechanism of quantized objects. So, what exactly does that imply? Mechanics is a branch of physics that deals with the interpretation of how things function in terms of forces, energy, and motion. And all quantized is simply represented in multiples of a small measurable unit. Then you would remember, just before the beginning of the century, classical physics or Newtonian mechanics is used to describe how stuff functions. And the fundamental characteristic of Newtonian mechanics is that anything was continuous; objects moved seamlessly across space; energy might arrive in an unlimited number of amounts; light rippled as a continuous wave, and there was no such thing as a minimum volume of anything.

Much of that shifted as quantum mechanics came into play. Energy, force, light, and motion were all quantized at this stage. You cannot only get some number for anything to be quantized; you have to have multiples of those minimum numbers. All of the properties of subatomic objects, and by implication, all stuff, is explicitly quantifiable according to quantum mechanics. Nature seemed grainy or jerky, hopping from one quantum number to the next without ever traversing the space in between. This creates a sense of unease over what is happening between such quantum states, or quantum jumps, to use a popular expression. In reality, it turns out that determine the accurate states of stuff among quantum states is impossible. There is no switch between quantum states in our interpretation (and this is a major element). You may have one, two, or three units of electricity, power, force, light, matter, or something, so you can't have two and a quarter or three and a half units.

This can be shown by an example. Assume a kid is hopping up the stairwell. A quantum leap is a move from one condition to another, similar to a kid leaping from one step to the other. She may even hop up to the fourth, or fifth stair, based on how much energy she had. She can't, though, hop between the stairwells and land safely.

Everything occurs in quantum jumps in the quantum universe, which is the system we exist in. The most critical argument I want to make is that all this falls into sharp focus when we try to weigh objects as fragile as atoms and explain them using Newtonian terminology.

Physicists have been debating the real existence of quantum leaps for over a century. There is now a solution, and in a real quantum fashion, everybody was partially right.

The expression "quantum leap" has taken a beating in recent decades; for certain viewers, it conjures up images of a stereotype for massive transformation or a sci-fi TV show featuring time machines. It explains one of quantum mechanics' central tenets: atoms possess energy levels that are discrete, and that electrons inside an atom will leap from an energy level to another but can't be detected between them.

Giants of physics such as Niels Bohr, who proposed the theory in 1913, Albert Einstein and Erwin Schrödinger fought over the details of these quantum jumps, namely if they were timing was spontaneous or the timing was instantaneous.

Yale University's Zlatko Minev and his team have now put an end to the discussion. "When we focus into a fine size, the leap is nor instantaneous nor as completely spontaneous as we assumed," Minev explains.

The researchers accomplished this by creating a quantum electrical circuit that super conducts 3 energy levels: the ground state, the default setting of an atom, a "bright" state linked to the ground state, and a "dark" state upon which the atom can leap.

To pump energy to the framework, they launched microwave beams at the synthetic atom. The atom was generally bouncing back and forth between the ground and bright states, releasing a photon each time it went from bright to ground. The molecule, on the other hand, would enter the dark state as it received a high-energy photon from its laser. The atom will remain in the dark state for long without releasing any photons because it was more robust than the bright setting.

By watching for a pulse of light of the bright state accompanied by a delay when the atom jumped into the dark phase, the physicists became able to say whether a quantum leap had begun. Minev relates it to forecasting the explosion of a volcano. "It's a spontaneous occurrence; none of us can guess whenever the next volcanic explosion will occur; but, there are some signs in the earth that we can track and use as an alert until the next eruption occurs," he states.

The atom's pause in light is analogous to earthquake alarm signals. It is difficult to guess when the next leap will happen in longer periods, as Bohr believed, but it is feasible on period's timescales of only a few milliseconds.

"The observation that this quantum leap was observed in a superconducting system instead of an atom indicates that we can manipulate this superconducting system in respects that we can't control normal atoms," says MIT professor William Oliver. He believes that in the future, we will be ready to do the very same things for actual atoms.

This control enabled the group to accomplish something that must have seemed difficult to Bohr: regulating a quantum jump.

If the physicists reached the atom via an electrical pulse shortly after the leap began, they might catch it and return it to the lowest energy – something that wouldn't have been feasible if quantum jumps were completely spontaneous and unpredictable. Instead, they discovered that the jumps still followed the same direction between the 2 energy stages, making it simple to foresee how to return them.

This demonstrates that quantum jumps aren't instantaneous, as Erwin Schrödinger believed – they require about 4 microseconds. "In certain ways, the leaps aren't leaps," Minev explains. "If you focus on these finer aspects, you will achieve something that you would not have imagined you might because of these small predictability windows."

According to Minev, this may one day be used to fix quantum computing errors. An unexplained quantum jump might indicate a measurement error, and this approach might enable researchers to detect the leap's beginning and compensate for the fault, or even undo it in the middle of the jump. "It's a very interesting scientific finding, and its importance to potential quantum computers will be determined by what quantum computing looks like," Oliver adds.

Chapter 9- Quantum Physics and Health

When you read the term "quantum," you might think about physicists experimenting on a revolutionary new idea. You have heard of quantum computing and how it could revolutionize the planet. Medication, a lesser-known sector, is beginning to enjoy the advantages of the quantum world.

A host of quantum innovations are being implemented in Europe as a result of the Quantum Technology Flagship an initiative of the European Union to improve a range of fields. Medicine, in fact, seems to be on the rise, with many programs now ongoing to see if we can develop diagnostic imaging to more effectively diagnose diseases.

- Quantum simulation may assist in keeping flights on schedule.
- Quantum technology
- Quantum cryptography: a double-edge sword
- In the not-too-distant future, quantum computers will outperform classical devices.'

Image: IBM Q quantum computer

One of these programs is macQsimal, which is developing quantum-enabled gyroscopes, magnetometers, atomic clocks, and more accurate gas concentration and electromagnetic radiation measurements utilizing tiny instruments such as quantum sensors. The venture, which launched in October 2018, aims to introduce some of the very first quantum-enabled technology to customers.

'The aim is to bring goods on the market as prototypes,' stated Dr. Jacques, program manager for macQsimal. 'At the end of the day, we want to be willing to move these devices to the next level and commercialize them.' However, we must consider the new wave of quantum sensors, which would use more sophisticated quantum effects such as entanglement and superposition.'

Sensors based on quantum mechanics

A quantum sensor is basically a very small computer, about the scale of a die, which can make extremely precise measurements by exploiting the quantum world's recognized strangeness. Entanglement occurs where particles are joined as one over long distances, and superposition occurs where particles exist in two locations at the same time.

This may be particularly helpful in areas such as brain imaging. Magnetoencephalography (MEG) scans currently depend on large, heavy machinery that must be kept cool with liquid helium or liquid nitrogen. As a consequence, the devices are not only massive, but they also can't get close enough to a person's skull to track brain function, so they have to rely on sensors to do so.

Dr. Haesler elaborated, "The aim is to substitute these devices with something of a helmet in which you will put all of the sensors, which you would put on the head, so you can increase the measuring precision." 'After that, you will build a helmet using hundreds of sensors.' So you can calculate where the magnetic force is originating from at hundreds of various points on the head.'

With the magnetometers it is making, the macQsimal group aims to demonstrate that this is possible. It may be possible to diagnose illnesses in a person's brain even more quickly if the machinery is greatly reduced in scale. It is hoped that the technologies they are building would be commercially viable in five years.

Such advantages may include medical imaging – capturing pictures of the heart to screen for diseases – which could profit tremendously from these small and more precise sensors, and also drug development – discovering potential medications to treat specific diseases. 'Most likely, there are a lot of options in the medical industry,' Dr. Haesler said.

Quantum properties including entanglement can be seen in new quantum sensors.

Hyperpolarization

Other scientists are looking at a quantum method called hyperpolarization to see how MRI scanners can be rendered even more sensitive and precise than they were before. This is the objective of the MetaboliQS initiative, which began in October of 2018.

The project manager, Dr. Christoph said, "We are essentially trying to render MRI perhaps a scale of 10,000 more responsive." 'By hyperpolarizing injected biomolecules, these molecules may be tuned to collect in a specific tissue.' And as they build up, the MRI would be able to sense what's going on more easily.'

To see what is really going on within our bodies, hyperpolarized MRI takes pictures by staring at the tiny cells. This is accomplished by utilizing specific biomarker molecules, which must first be cooled to -270°C before being heated to body temp. This procedure not only requires a long time (at minimum 30 mins.), but it also costs a lot of money.

The MetaboliQS group thinks that by utilizing diamond-based quantum sensors, they would be able to carry out the whole procedure with either moderate cooling or no chilling at all. This can make it easier for MRI machines to detect time-sensitive results in the body, like cancer cells, as well as capture more precise photographs.

'If you refine the photographs, you see even more information, and you can tell the difference between early-stage disease and later-stage disease, or dead tissue,' Prof Nebel said. 'Having clearer photos ensures you have a clearer knowledge of medicine.'

This could lead to new applications for MRI scanners, like implant testing or further understanding of how diseases evolve in the organisms. If effective, MRI imaging may be one of the first fields of medicine to profit from quantum technologies.

Prof Nebel stated, "Hyperpolarization is certainly something that could be the first possible (medical) use of quantum science."

'Having clearer photos indicates you have a better knowledge of medicine.'

Health Conditions

If these programs are good, they would be able to solve a wide variety of conditions. Using the support of more precise MRI scanners, Alzheimer's disease and dementia could both be given a diagnosis more efficiently, according to Dr. Haesler. Furthermore, cardiac and brain scans will be advantageous, enabling other problems to be seen in greater depth.

'You can sense new neuron behavior very nicely with these quantum detectors we are currently building,' said Dr. Nebel. 'Essentially, we should look at very tiny molecules and Biosystems.' On an atomic scale, this is essentially MRI.'

The next move would be to get these goods to market to demonstrate their viability. And it's anticipated that, with the aid of the EU's quantum flagship initiative, inventions like these will usher in a fascinating new quantum age that will have a direct effect on our lives.

That isn't only exclusive to the medical field. The software is now looking at ways to upgrade atomic clocks and other technologies that might help us improve our cell networks, for instance. Medicine applications, on the other hand, are expected to come first, with significant consequences for our wellbeing.

We believe the magnetometer and atomic clock will be available in 4 years,' told Dr. Haesler. 'We're already focusing on the 2nd gen of detectors, which will be more adaptive and available in 10 to 15 years,' says the researcher.

No needles

The York University has developed a pad that can be added to the skin to administer targeted treatments without the use of hypodermic needles. Nanject is a patch that can distribute cancer therapies without damaging healthy cells.

The following is how it operates: Antigens (additives that react to antibodies) are coated on the nanoparticles until they are inserted into the bloodstream, where they adhere to cancer cells. The patient is then treated in an MRI system, which causes the molecules to warm up and kill cancer cells. The molecules cool off as the unit is switched off and may be withdrawn from the individual without harming the user.

The Nanject patch substitutes a single syringe with several tiny ones constructed of polymer nanofilaments that inject the drug into hair follicles, which might be appealing to needle-phobic patients.

However, there is another, even more significant, the advantage of using nanotechnology for drug delivery: It eliminates some of the most difficult obstacles to drug distribution, especially in rural and distressed areas. A patch eliminates the need for a qualified nurse or specialist to deliver drugs so they can be self-administered using a procedure as straightforward as applying a Band-Aid. Because nanoparticles are not digested by gastric acid like pill drugs, nanotech drug distribution allows for smaller doses. Finally, procedures like the Nanject may help deter disease transmission by unsterilized needles, which is a big issue in underdeveloped countries.

Human Biology Hacking

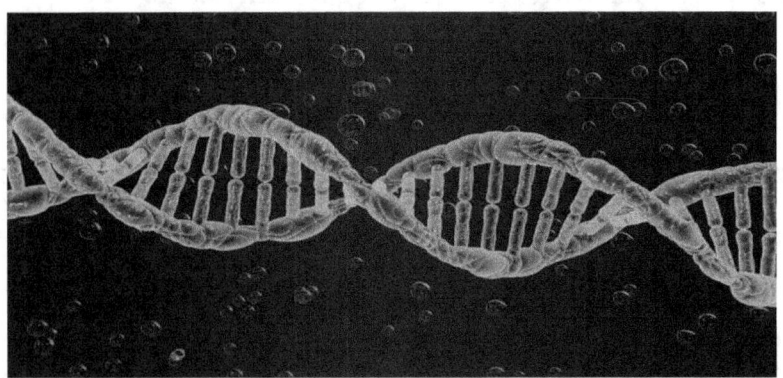

Quantum mechanics has the ability to give us more knowledge regarding human genetics, in addition to better cancer detection and needle-free therapies.

We will more easily sequence DNA and address other Big Data challenges in medical services using quantum computing.

Using a novel method of laser microscopy based on quantum mechanics, Australian scientists recently found a way to investigate the internal dynamics of a living organism. We can also sequence DNA and address other Big Data challenges of health care more easily with quantum computers. This opens the door to tailored treatment focused on a person's genetic makeup.

I hope that, having arrived at this point, you have had a positive reading allowing you to transcribe all your positive reflections in a review that will be of fundamental help to me, thanks for choosing my work.

Conclusion

"Because I can't identify the real issue, I believe there isn't one, but I am not certain there isn't."

The great physicist, Feynman said this regarding the popular mysteries and confusions of quantum physics, the idea scientists use to explain the tiniest particles in the World.

Some scientists believe we already know what conscience is or if it is a figment of our imagination. Many people, on the other hand, believe we really don't understand where consciousness emerges from.

Some theorists have also used quantum mechanics to illustrate the perennial mystery of awareness. The concept has often been greeted with cynicism, which is understandable given that explaining one puzzle with another doesn't seem appropriate. However, those concepts are neither necessarily absurd nor subjective.

For one aspect, the consciousness appeared to push its entry into initial quantum theory, much to the displeasure of physicists. Furthermore, quantum computers are expected to be capable of doing something that conventional computers can't, reminding us about how our minds can do tasks that AI can't. "Quantum consciousness" is commonly dismissed as mystical nonsense, but it keeps popping up.

Quantum physics is the most accurate explanation we have for understanding the universe at the sub-atomic and atomic stages. The idea that the result of a quantum test will vary based on whether or not we want to calculate any value of the objects concerned is the most well-known of its puzzles.

The early founders of quantum physics were profoundly disturbed when they first observed this "observer influence." It appeared to be undermining the fundamental premise of all scientific knowledge: that it's an empirical universe out there that exists independently of us. What does "reality" actually imply if the way the universe acts relies on how – or even if – we gaze at it?

The "double-slit" is the most common example of the mind's interference with quantum mechanics.

Scientists feel compelled to believe that objectivity was a figment of their imagination, and that conscience must be given a voice in quantum mechanics. That didn't make sense to others. The Sun, surely, does not work just when we gaze at it, as Einstein once lamented.

Some scientists now believe that, whether or not awareness has an effect on quantum physics, it might have arisen as a result of it. Quantum mechanics, they believe, could be needed to truly comprehend how the brain functions.

May it be whether, just when quantum phenomena seem to be in two positions at once, a quantum mind may simultaneously contain two internally exclusive ideas?

These theories are theoretical, and it's possible that quantum mechanics plays little intrinsic involvement in the mind's functioning. These ideas, if nothing else, demonstrate how oddly quantum theory causes one to wonder.

Pascual Jordan, a physicist who collaborated with quantum visionary Niels Bohr, put it this way: "Observations don't only disrupt what needs to be monitored; they often create it... [A quantum atom] is compelled to take a certain role." "We generate the effects of measurement ourselves," Jordan stated.

If this is the case, empirical reality seems to be thrown out the window.

It gets stranger from there.

We might attempt to manipulate nature into exposing its hand if it appears to change its behavior based on whether we "observe" or not. We may do this by measuring the direction a particle followed across the slits after it had passed through them, only once it had progressed through them. It should have "made the decision" whether to follow either or both paths by then.

The collapse may be induced by the simple act of observing rather than for any physical disruptions caused by measurement.

In 1970, John Wheeler suggested a method for doing so, and the "delayed decision" experiment was carried out the following decade as discussed in the book. It makes observations on the pathways of subatomic particles after they could have taken one direction or a superposition of 2 paths using clever techniques.

It appears that it doesn't make a difference whether we prolong the calculation or not, much like Bohr confidently expected. We lose any disturbance as long as we calculate the photon's direction until it is actually detected at a detector.

It's as if they "know" not only if we're looking, but also if we will look in the future...

When we uncover a quantum photon's course in these tests, the cluster of potential paths "collapses" into some kind of single defined state. Furthermore, the delayed-choice investigation indicates that the breakdown may be triggered by the simple act of observing rather than some physical interference caused by measurement. But would that imply that actual collapse occurs only because the outcome of a calculation has an effect on our conscious awareness?

It's difficult to escape the conclusion that quantum mechanics and consciousness are linked in some way.

Wheeler also considered the possibility that the existence of living creatures capable of "observing" has turned a plethora of potential quantum histories into a single concrete background. In this way, we become partners in the Universe's development from the beginning, according to Wheeler. We exist in a "participatory world," as he puts it.

Physicists really can't settle on the right approach to view these quantum phenomena, but your interpretation is (for the time being) open to you.

Roger Penrose, a British scientist, proposed in the 1980s, discussed in previous chapters that the relation might operate in both directions. He speculated that whether consciousness may influence quantum physics or not, quantum physics might be active in consciousness.

Maybe there are structural frameworks in our minds that can change their condition in reaction to an individual quantum phenomenon, Penrose wondered? Couldn't these systems, like the photons in the double-slit, follow a superposition? Could those quantum superpositions then manifest themselves in how neurons are activated to communicate by electrical signals?

Perhaps our capacity to maintain apparently contradictory mind states isn't a quirk in thought, but rather a true quantum influence, according to Penrose.

It's possible that quantum physics plays a role in consciousness.

Overall, the human mind seems to be capable of processing functions that greatly outstrip those of modern machines. Perhaps we would be able to perform computing functions that are impractical on traditional computers.

With the help of physician Stuart Hameroff, Penrose expanded on this definition. He proposed that the frameworks working in quantum cognition may be strands of protein named microtubules. Many of our cells, like the neurons in the brain, include these. Microtubule movements, according to them, will follow quantum superposition.

However, there is no proof that this is even slightly possible.

It has been proposed that results published in 2013 endorse the theory of quantum superpositions in microtubules, however, those research did not discuss quantum consequences.

Quantum impacts of living organisms have been discovered by other investigators.

Because of a mechanism is known as decoherence, quantum phenomena such as superposition are readily eliminated. This is induced by a quantum entity's encounters with its surroundings, which causes the "quantumness" to leak out.

Electrical pulses are generated by the passing of atoms that are charged electrically through the boundaries of nerve cells, resulting in nerve signals. If any of these molecules were in superposition and clashed with a neuron, the superposition would disappear in an instant of a billionth of a sec... A neuron would take at least a hundred thousand trillion times longer to send a signal.

As a consequence, theories concerning quantum phenomena in the mind are widely dismissed.

Most people believe that awareness and the brain, and possibly vice versa, should be left out of quantum mechanics. And besides, we have little idea what awareness is, far less a hypothesis to explain it.

We all understand what red looks like, but we don't know how to describe it.

However, that notwithstanding, the concept has a long past. It's been devilishly difficult to get rid of the "observer influence" and the mind since they first infiltrated quantum physics in the early years. Some experts think we would never be able to do so.

Among the most esteemed "quantum theorists," Adrian Kent of Cambridge University, hypothesized in 2016 that conscience could modify the actions of quantum particles in subtle yet observable ways.

Kent is apprehensive about the proposal. "There is no convincing justification to conclude that quantum theory is the best theory with which to attempt to develop a concept of consciousness and that quantum theory's challenges may have something to do with the issue of conscious awareness," he acknowledges.

Any line of reasoning about the connection between consciousness and physics is futile.

However, he claims that it's difficult to see how a definition of consciousness founded solely on pre-quantum mechanics would account for any of its characteristics.

Why our aware brains may feel unique stimuli like the color blue or the scent of flowers a highly perplexing mystery. We all understand what blue looks like, with the possibility of those with vision problems, but we do have no means of communicating the feeling because there is little in physics that shows us what it could be like.

This form of sensation is referred to as "qualia." They seem to us as integrated properties of the outer universe, but they are really products of our minds, which is difficult to understand. Indeed, it was called "the hard challenge" of consciousness by theorist David Chalmers in 1995.

This has led him to say that "we might make significant headway on identifying the situation of consciousness development if we assumed that consciousness change quantum probability (though somewhat only marginally and subtly)."

In this perspective, it does not precisely decide "what is reality" However, it can have an effect on the likelihood that each of the potential actualities allowed by quantum physics is the one we experience, in a manner that quantum mechanics cannot anticipate. We should search for certain results experimentally, according to Kent.

He also calculates the likelihood of discovering them. "I will put a 15% probability on something directly related to consciousness causing quantum theory anomalies, with a 3% chance that this would be experimentally observable over the next 40 years," he states.

If this occurs, it can fundamentally alter our understanding of both physics and the brain. That seems to be a possibility worth pursuing.

www.ingramcontent.com/pod-product-compliance
Lightning Source LLC
Chambersburg PA
CBHW071412210526
45465CB00001B/357